Race, Monogamy, and Other Lies
They Told You

The publisher gratefully acknowledges the generous support of the General Endowment Fund of the University of California Press Foundation.

, and
Other Lies They Told You

Busting Myths about Human Nature

Agustín Fuentes

UNIVERSITY OF CALIFORNIA PRESS

Berkeley · Los Angeles · London

University of California Press, one of the most
distinguished university presses in the United States,
enriches lives around the world by advancing
scholarship in the humanities, social sciences, and
natural sciences. Its activities are supported by the UC
Press Foundation and by philanthropic contributions
from individuals and institutions. For more
information, visit www.ucpress.edu.

University of California Press
Berkeley and Los Angeles, California

University of California Press, Ltd.
London, England

Library of Congress Cataloging-in-Publication Data

Fuentes, Agustín.
    Race, monogamy, and other lies they told you :
Busting myths about human nature / Agustín Fuentes.
        p.   cm.
    Includes bibliographical references and index.
    ISBN 978-0-520-26971-2 (cloth : alk. paper)
    1. Behavioral evolution.   2. Human behavior.
3. Human evolution.   4. Social evolution.   5. Physical
anthropology.   I. Title.
    BF 698.95. F85   2012
    155.7—dc23

                                    2011049949

Manufactured in the United States of America

21   20   19   18   17   16   15   14   13   12
10   9   8   7   6   5   4   3   2   1

In keeping with a commitment to support
environmentally responsible and sustainable printing
practices, UC Press has printed this book on Rolland
Enviro100, a 100% post-consumer fiber paper that
is FSC certified, deinked, processed chlorine-free,
and manufactured with renewable biogas energy.
It is acid-free and EcoLogo certified.

Contents

Illustrations

Preface

Three major myths—about race, aggression, and sex—have a negative impact on our society and inhibit an accurate understanding of what it means to be human. These myths create a false set of societally accepted "truths" that in turn cause a range of problems for us. The myth that humans are divided into biological races—that black, white, Asian, etc. are natural categories—helps generate and maintain intolerance and inequality, and leads to difficulties in creating and sustaining communities in our increasingly diverse society. The myth that removing the constraints of culture and civilization reveals the innate, violent beast within us (especially in men) restricts how we can relate to one another, encourages fear, and enables an acceptance of certain kinds of abuse and violence as natural or inevitable. The myth that men and women are dramatically different in behavior, desires, and perspectives due to natural differences in "internal wiring" facilitates poor intersexual relations, creates and maintains sexual inequality, and causes a range of problems for individual men and women laboring under a preconception about who and how they are supposed to be.

Busting myths of human nature is not like busting the myth that a tooth left in Coca-Cola overnight will dissolve or that humans only use 10 percent of their brains. Most false beliefs are clearly refutable with a single, usually simple, test. There is not going to be one, or even a few, simple tests that will destroy every piece of the myths about race, aggression, and sex, but we can still show that they are wrong. There are information, data, and concepts out there that will demonstrate that these myths are false.

Busting myths about human nature requires some effort. It means breaking the stranglehold of simplicity in our view of what is natural and forcing ourselves to realize that being human is very complicated. It means challenging common sense and our reliance on generalities and popular perception, and actually delving into the gritty details of what we know humans actually do. This is the goal of the book and, hopefully, by the time you get to the last page, you will agree that the major myths about race, aggression, and sex are neither correct nor do they explain core parts of human nature.

To get you to that place I have a plan. At some points in the book it might not seem like it, but rest assured, I do. It turns out that to be able to bust these myths we need to have a shared baseline of understanding, a starting point of knowledge from which to tackle the myths themselves. This baseline includes what we mean by "myth" and "human nature" and why these two things are so important in our society. It also includes what culture, evolution, and genes are and how they influence you and me and everyone else on the planet. The biggest challenge is to present this information in a way that covers enough concepts and details and yet boils them down into a few salient points: a basic myth-busting tool kit. The first three chapters of the book are this tool kit and the set-up for the real work of busting myths about human nature that occurs in chapters 4–6. The final chapter provides a set of take-home points and some concluding thoughts. There is also an appendix that serves as a quick primer on how to bust myths about human nature yourself.

WHY SHOULD YOU BELIEVE WHAT I HAVE TO SAY?

Actually, I don't want you to believe me, but I do expect that after reading this book you'll be in a better position to make up your mind for yourself on the main themes it covers. Understanding who is making statements about human nature and how to assess where they derive their expertise is critically important. Because I am writing this book, selecting the information, presenting it in a certain way, and trying to lead you to a set of conclusions, you need to know a bit about me to be able to judge the validity of my perspective.

I am a scientist, a specific kind of social scientist called an anthropologist, a specialist in human and primate behavior and evolution. I am also a teacher, an active researcher, and was trained in anthropology, biology, and evolutionary theory by some of the most respected professors and researchers in those fields. Having a bit of background on who I am and where I come from provides context for how I conceived this book and, hopefully, will make you feel more secure that the information here is reliable, that I am well qualified to review this information, and that my conclusions as to what it might mean are reasonable.

I was born in the United States to a Spanish father and American mother, both educators (university and primary school). Although the majority of my schooling has been in the United States, I have also lived in Spain and Indonesia. I speak English, Indonesian, and Spanish well, and a few other languages passably. In the United States I have lived primarily in California, Indiana, Texas, and Washington, and I have close relatives in California and New York as well as in Madrid (Spain). I have conducted research in the United States, Southern Spain (Gibraltar), Singapore, and many locations in Indonesia. I've traveled in Morocco, Micronesia, Thailand, Malaysia, China, Japan, Korea, Australia, New Zealand, Central America, Venezuela, Mexico, Canada, and most of Western Europe. I mention these pieces of biographical information because they are relevant to understanding my personal view of humanity: I come from a family of educators and I travel a lot. This means that I am lucky enough to be able to see how people look, live, and think across much of the globe, and as an anthropologist I am always considering how their lives and mine intersect and differ. I am truly amazed at how many similarities, and differences, go into making us all human. This experience has led me to have an open mind to opinions and beliefs as I realize that my personal experience is only the tiniest fraction of all the experiences that humans have and that I have to be ready to listen to others as they might know much that I do not.

I received all of my university degrees from the University of California at Berkeley. My bachelor's degree is a double major (five years) in anthropology and zoology, and my master's and PhD degrees are in anthropology. I was lucky enough to be able to take a wide range of classes at Berkeley with many amazing professors. As an undergraduate I was in the last graduating class of the Department of Zoology (now part of Integrative Biology) where I benefited from an amazing focus on natural history, the idea that you need to see organisms, watch them in their daily lives, and get a strong idea of *what* things they actually do before you make a series of assumptions about *why* they do what they do. This melded perfectly with my anthropology major, where this focus on observation and careful and meticulous examining of the world around us and the humans in it was the core principle in my training. Both as an advanced undergraduate and a graduate student I was able to take courses and seminars with some of the top evolutionary theorists, biologists, and anthropologists at Berkeley (and occasionally at UC Davis as well).

This was, and remains, a humbling experience. It demonstrated to me how much time and effort it takes to know even small amounts of specialized knowledge and that most questions can be approached from multiple areas of specialization, not always generating the same answer.

After receiving my PhD in December 1994, I taught at Berkeley in the Department of Anthropology for a few years, and then went to Central Washington University to help enhance their anthropology program and design and direct an interdisciplinary program in primate behavior and ecology (a collaboration between the anthropology, biology, and psychology departments). In 2002 I moved to the University of Notre Dame, where I am currently a professor of anthropology. In this time period I have taught thousands of students, run many research projects and field schools, received generous grant support, and published a large number of peer-reviewed articles, book chapters, and books.

My research projects range from examining primate and human behavior and ecology to modeling patterns in human evolution to examining the roles of pair bonding and sexual behavior in humans and other primates. I have investigated the role of interspecies encounters and the patterns of disease transfer that sometimes accompany them. Recently, I've been particularly focused on the interactions between humans and other animals (especially monkeys and dogs) and trying to understand what it is about relationships between people that is particularly special in our evolutionary histories. Some of my major findings have helped to redefine what pair bonding and monogamy mean for primates and humans, clarified the importance and complexity of cross-species relationships, described how modern evolutionary theory helps us better understand our behavior, and illustrated how cooperation between members of a group, and even between groups, can be a backbone for understanding how humans became so successful over our evolutionary history.

This life history, training, and research experience has influenced who I am and how I think about topics related to being human. I am not an essentialist or biological determinist. That is, I do not believe that there are strict set-in-stone patterns at the core of human behavior. Nor am I a true social constructivist, thinking that humans emerge as a blank slate and that experience and social pressures alone shape our behavior. My experience leads me to believe that both of those perspectives are too simple and too limiting to explain humanity. Influenced by my travels and interaction with peoples across the planet, I am probably considered relatively liberal politically (in the United States), but I do not identify with liberal American positions across the board by any stretch of the imagination. Most importantly, I am always ready to be wrong. I, perhaps naively, still believe that the quest for knowledge, for explanations, is best done via a method (such as science) where one's hypotheses are testable and the goal is to prove one's hypotheses wrong, or support them. My most memorable and effective discoveries have always been the accumulation and analyses of new information, which showed previous conclusions to be wrong. We can never really prove ourselves right, only narrow the range of possible truths down to a few very probable truths.

In my research, teaching, and life experience I have now accumulated enough information to convince myself that we can get much closer than we currently are to probable truths about being human. As an educator and researcher I feel the need to share this perspective outside of the classroom and laboratory. As an anthropologist I see the need to correct misinformation or lack of information about these topics. And as a member of our society I feel the urge to bust these myths because of the way they constrain our thinking about, and being with, each other.

Acknowledgments

The researching and writing of this book have taken many years and would not have been possible without the support, assistance, and intellectual contributions of many, many people. I particularly want to thank my amazing colleagues at the University of Notre Dame in the anthropology department and beyond, and my close colleagues in the worlds of anthropology, biology, and psychology. There are too many individuals in these categories to list here, but so many of them deeply influenced my thinking, practice, and research. I also must thank my students over the last seventeen years for pushing me intellectually and pedagogically. Without the insights and engagement of all these people I would not have been able to write this book. I also thank the University of Notre Dame Institute for Scholarship in the Liberal Arts and the Dean of the College of Arts and Letters for specific support on this project.

I am indebted to Blake Edgar (sponsoring editor) for his support and encouragement in the initial development of this book project and for his comments and suggestions throughout, and to the production staff at University of California Press, Rachel Berchten (project editor) and Kate Marshall (associate editor), and to the stellar copyeditor, Kathleen MacDougall, for making the manuscript into a book. I also thank Alexander Trotter for creating the index. The book you are reading was made much better by the excellent critical reviews of earlier versions by Karen Strier, Robert Sussman, and one anonymous reviewer. Specific chapters were greatly enriched by intellectual engagement with Jim

McKenna, Katie C. MacKinnon, Nick Malone, Hope Hollocher, John Archer, Susan Blum, Rahul Oka, Mark Hauser (the archeologist), Daniel Lende, Greg Downey, Douglas Fry, Julienne Rutherford, and others, as well as my presentations for the Hesburgh lecture series sponsored by the University of Notre Dame alumni association. Any and all omissions or errors in this book are entirely my fault.

Finally, I thank my family. My parents and their partners, aunts and uncles, all of my siblings, nieces and nephews, parents-in-law, and cousins across the globe for their unabashed support and the ever-present discussions on the topics in this book. I especially thank my partner, Devi Snively, for helping me develop these ideas, for partaking in (and putting up with) the life of the academic, and for sharing in, and shaping, my passion for the power of words.

PART ONE

Myth-Busting Tool Kit

1

Myths about Human Nature Are Powerful—and Misleading

There is a shared set of beliefs about human nature that shapes the way we see the world—common assumptions about race, aggression, and sex that are seen as just part of being human.

While we might not always admit it in public, most people think that there is a specific set of biological differences between various kinds of people in the world, and that if you strip away society and laws, humans become beasts, with survival of the fittest and the bigger, badder, more aggressive taking control. And of course, nearly everyone knows that it is natural that men and women want, and need, different things from sex and personal relationships.

These beliefs are myths based on misinformation, partial truths, and a large dose of ignorance as to what we actually know about our species. This book is focused on challenging what many people assume is common knowledge about what it means to be human. We are going to use information from a wide range of researchers and research projects to bust these myths and replace them with more accurate stories about who we are and what we do.

Why do these concepts about race, sex, and aggression seem to be common sense to so many people? It is largely because of the shared assumption that under the thin veneer of culture we have a basic set of instincts, a raw humanity. There is a popular perception of what human nature is, and common views of race, aggression, and sex permeate society. This can be encapsulated in three key myths:

1. Race: Humans are divided into biological races (black, white, Asian, etc.).
2. Aggression: Removing cultural constraints reveals the violent beast within us (especially in men).
3. Sex: Men and women are truly different in behavior, desires, and internal wiring.

By the end of this book you will see that what we know about these topics demonstrates, unequivocally, that humans are not more naturally monogamous, aggressive, and violent than we are polygamous, peaceful, and egalitarian, that men and women are not nearly as different as one might think, and that even though humans all belong to one race, racism matters. Being human is a lot more complicated than many of us think, but myths about human nature are powerful and remain quite popular.

WHAT IS A MYTH?

If common sense is as much an interpretation of the immediacies of experience, a gloss on them, as are myth, painting, epistemology, or whatever, then it is, like them, historically constructed and, like them, subjected to historically defined standards of judgment. It can be questioned, disputed, affirmed, developed, formalized, contemplated, even taught, and it can vary dramatically from one people to the next. It is, in short, a cultural system, though not usually a very tightly integrated one, and it rests on the same basis that any other such system rests; the conviction by those whose possession it is of its value and validity. Here, as elsewhere, things are what you make of them.
—Clifford Geertz (anthropologist)[1]

In this book we are interested in myths as stories, or explanations, of why things are the way we think they are. They make up a part of what many of us would call common sense: the stuff that you just know about the world around you, especially about race, sex, and aggression. This is why they are so powerful. By helping us make sense of the behaviors we see around us and the symbols we use, they allow us to go on from day to day, appearing to understand our world without having to reanalyze, or critically analyze, every day's situations.

For example, if someone makes a joke about women and shopping or a man reacts violently to a sports event, you already have a baseline of explanation in your head that allows you to "get" the joke (because shopping is part of being female) or understand the man's response (because men "get all testosteroned out" over sports). Now, in both of these examples there is some societal truth: many women do like to shop and many men do get aggressive about sporting events. However, are these things actually part of our nature or is something more interesting going on?

More subtly, in our society most people rely on a set of assumptions about someone when they see them, or first meet them, based on which race they appear to be. It's not that we are naturally inclined to be racist, or even racial, but rather that race means something in our society and we have a whole suite of myths about what to expect and understand about people and races.

None of these reactions are necessarily conscious thoughts; rather, the myths are so pervasive that these responses often go on without any active consideration on our part. The myths provide explanations and contexts so that we don't have to: they supply ready-made common sense. This does not mean that everything about our societal myths is untrue or that all such myths are false. There are many myths that have a lot of accuracy; however, the myths about race, aggression, and sex generally do not, or at least not in the ways we tend to think they do.

Dictionaries tell us that the word "myth" is a noun and define it as a traditional story concerning the early history of a people or explaining a natural or social phenomenon, typically involving the supernatural, a widely held but false belief, or a fictitious person or thing. The principal definition tells us a myth is a popular but false way of explaining things. According to the philosopher Mary Midgely, "we are accustomed to think of myths as the opposite of science. But, in fact, they are a central part of it, the part that decides its significance in our lives. So we very much need to understand them. . . . They are imaginative patterns, networks of powerful symbols that suggest particular ways of interpreting the world."[2]

We usually differentiate information associated with science from other types of information. However, what we think of as scientific realities are often filled with myth. For example, scientists of the 1700s were convinced that humors (liquids in the body) could move around and change the body as needed and so the medical establishment treated patients "scientifically" with that myth as their starting point. Now we know that blood does move through the body and affects the health and status of the body, but it does not do so in the ways that doctors in the 1700s thought it did. Some aspect of reality (the circulation of blood) and a major component of myth (the power of the humors) worked together to create a baseline reality that was accepted until other, more accurate, information came along and was integrated into society's (and science's) myth structure. The myth of the humors has not left us totally. Think of how many times we use the term "bad blood" to refer to ill health or ill will between people, implying that the state of the blood (humor) is what is driving health and behavior.

This way of thinking about myths is a bit different from what many people mean by the word "myth," where in most cases the referent is presumed to be Greek and Roman myths, Native American myths, or broader religious and spiritual stories. However, there are many similarities. The Greek myths were explanations for natural phenomena. Take the myth of the Pillars of Hercules. If you go to the Strait of Gibraltar (the narrow strait between southern Spain and northern Morocco that links the Mediterranean Sea with the Atlantic Ocean), you see two amazing, granite mini-mountains rising just off the coastline, the Rock of Gibraltar to the north and Jebel Musa to the south. In one version of the myth Hercules has to cross a set of mountains and, rather than climb over them, he uses his terrific strength to move them apart, joining the two seas as a result. Here the myth explains a striking aspect of the local geology. Myths also acted as lessons, guidelines, and justifications for how one should live one's life. For example, the myth of Icarus (who flew too close to the sun with wings of wax despite his father's warnings) is a parable about respect and attention to parents, about caution in risk-taking, and about the lure of the beautiful and prohibited. Unlike these ancient Greek myths, the myths we are concerned with in this book are not about heroes, monsters, and mountains. Rather, they are the day-to-day beliefs we carry with us that act in the same way to explain, give reasons to, and help us navigate the world we come into contact with. These myths about human nature can be potentially harmful to us as a society. The mythical ideas we share about humanity can affect the ways in which we behave toward and think about other people and set up expectations and assumptions about where we are as a species. We have many beliefs about why humans do what they do; but a number of these beliefs, as I will point out in this book, are neither factual nor a true baseline for humanity.

Myths have an impact on the way we think and feel

Our societal myths help us navigate our daily lives by providing handy basic assumptions about the goings-on around us; they help move our day along, even if subconsciously. When a man screams out in anger from a car stuck in traffic on the freeway, or a woman cries after her grocery bags tear and the contents fall to the floor, we respond to what happened. But, at the same time, we also have a ready-made explanation for a man's rapid turn to aggression or violence and the women's emotional response. When we hear about a couple's breakup around infidelity, we tend to make assumptions about whom, what, and where, based on our preconceptions about males and females. When a group of high school kids lines up to pick sides for a basketball game, assumptions are made about the abilities of the potential players based on the color of their skin and their ethnic backgrounds. The same occurs when a teacher watches a classroom of mixed ethnicities and genders sit down to take a standardized exam. We have expectations about behavior and potential based on both our life experiences and our myths about humanity. Together, our prior experiences and our shared myths act to build common sense or provide basic explanations for the world we live in and help shape that world and our behavior in it. Let's use two very simple examples to demonstrate these points: one from a myth we'll bust later in the book and another from a very popular set of myths about health, travel, and cures.

It is commonly assumed that men are loath to ask for directions. This is the brunt of many jokes that persist because we are participants in the myth about who men are. However, the myth is not really about asking questions, it is about how we define and understand male biology and male nature. Inherent in this popular perception about men not asking for directions are some assumptions about male gender: men are proud, men like to be do-it-yourselfers, and it is masculine to be in charge and know where you are going. These are important components of the gender-role definition for males in our culture (indeed, in many cultures). So at one level, the joke about men not asking for directions rests on a set of cultural expectations about how males should act, but this is not the myth. The myth is what underlies much of these cultural assumptions, the part that most people do not actively think about when laughing at the jokes about men and directions.

What we are really interested in here is the myth of male nature that creates an evolutionary, or biological, story to support cultural expectations of male gender. This myth involves the assumption that men have better spatial reasoning abilities than women, including innate mathematical abilities. This makes men more likely to be able to navigate spatial problems (like getting from one place to another) by individual actions such as map reading, calculating distances, imagining complex spatial layouts, and then actually following them. Now, that men may have this superior spatial capability is not totally accurate or inaccurate (as is examined further in chapter 6), but even this is not the core of the myth. The real meat of interest here is our mythical explanation for *why* men might have these spatial abilities over women: man the hunter.

We share a mythical notion about men as hunters. Most people would agree that in our past, humans relied heavily on hunting animals for food and hides and bones. Most would also agree that this hunting was done by men and not women. If this were the case, over time men would have become really good (biologically), and better than women, at the skills needed for hunting: spatial reasoning, tracking game, mentally mapping landscapes, and hand-eye coordination for making and using tools and weapons. It turns out that for the vast majority of human history (that is, the last two million years or so) we do not have good evidence for who had these skills (nothing one way or another) even though most researchers make the assumption about men, hunting, tool use, and tool making.[3] What we do know is that in most of the few remaining hunter/gatherer groups left on the planet, men do the lion's share of the big game hunting (even if women bring home a large portion of the actual calories eaten by the group in the form of gathered foods). We also have evidence that over the last 10,000 years there has been an increasingly common pattern across human societies of big differences between male and female roles in the acquisition and processing of food.

So, despite the myth that men evolved as hunters and tool users and makers and women did something else (usually we think of them preparing the food and tending to babies), we don't have any evidence that early men made more tools than early women (or even that there were any differences in who made which tools), or that one gender had more spatial knowledge of the areas used by the group. We know that in societies across the planet today there are large differences in the types of tools men and women make and use, and that there are widening differences in the use of living and working space as agriculture, industrialization, and economic stratification increase. We also have no evidence indicating who prepared the food in the past, but we do know that today preparation of food varies across cultures, with a majority of societies having women do most of the preparation work. We also have widely varied results from tests that measure male and female math and spatial abilities (though actually there is very little difference overall: see chapter 6) as well as from tests that measure hand-eye coordination, although men seem to be able to throw things a little better and farther. Still, that pattern might also be related to men being bigger and having higher muscle density on average than women.[4]

There is hard and fast evidence that men and women today are different in some facets of hunting and that there are differences between modern male and female roles in regard to acquiring and preparing food. Is there sufficient evidence to support an assertion that over the history of the human species men (and not women) were the hunters and that this leads to a better, natural, innate, male ability at spatial reasoning and navigation? No, there is not. This myth comes from a mix of information about modern hunting and gathering societies, rooted in current cultural expectations of gender roles (how men and women are supposed to behave), and some obvious average differences in size and strength between males and females. This practice of making a large set of assumptions from a small bit of data and then asserting it as a "truth" about the natural world is common in many arenas of human behavior, especially when we are using these assumptions to think about the nature of humanity.[5] Critically thinking about our popular notion of men not asking for directions reveals the more serious and powerful myth about men's nature. Assessing that underlying myth shows a more complex reality than the one reflected in facile assumptions about men and women.

In a very different, but related vein, let's take a look at a set of beliefs that we will not be reviewing in depth in this book, but which gives us a good idea about how cultural myths can have financial and societal impact. There is a widespread assumption that traveling on planes can be dangerous because of the recirculated air and the frequency of sick passengers on board. Most people think that air on planes is largely recirculated, enabling germs to flow around the cabin and infect multiple people. People generally have a notion that when they travel by plane they run a higher risk of catching a cold than in other contexts (working in public buildings, traveling by train, etc.). Victoria Knight-McDowell (a schoolteacher) and her husband, Rider McDowell (a writer), developed a prophylactic (something you take to avoid catching something else) dietary supplement called Airborne. Airborne's initial packaging and marketing focused on the assumed risk of getting sick while flying. By 2008 this product was generating over $300 million in sales and could be found in travelers' pockets across the United States (including my academic colleagues and even members of my family). The product label implies that taking it regularly can boost one's immune system and thus prevent or cure colds (however, it never states that it actually does so). The ingredients include vitamin C, which has been shown to have limited success at reducing the length of a cold (largely by reducing the symptoms), and many people in our society think that taking a lot of vitamin C can help rid them of a cold. None of the other ingredients have been demonstrated to be effective against colds or specifically beneficial for the immune system (nor has Airborne itself, which is not regulated by the FDA).[6]

How is this anecdote relevant to our myths about human nature? It reflects a broader myth about biology and technology that influences behavior. Even a small myth, when popular, can affect the way a society thinks, acts, and spends money. How is it a myth? Well, for one, it is simply not true that air on planes is predominantly recirculated or germ-laden. Modern airplanes mix some compressed air with air drawn in from outside and the mix is about 50 percent at any given time. The air is refreshed throughout the flight with very effective filters and there is a total changeover in cabin air (that is total cabin air moving through the filters) every three minutes or so.[7] So the danger from planes and disease (unless one is seated directly next to someone who is highly contagious) is pretty minimal relative to what one risks in most large office buildings. Also, what we call the common cold comes from viruses (mostly a group of rhinoviruses and corona viruses), meaning that one would need an antiviral drug or compound to prevent them. We do not currently have one. There is no vaccine for these viruses and the only demonstrated method of avoiding a cold is to ensure the viruses do not get into your upper respiratory tract (washing hands regularly and not touching your mouth and nose is probably your best bet).

What we actually know about technology shows us that planes are not particularly dangerous places to catch a cold, and biological and medical research shows us that there is nothing (currently) that we could consume that would help us avoid or prevent catching a cold. Yet, tens of thousands of people have purchased Airborne prior to flying in the expectation that they are about to undertake a disease risk (the plane journey) and that these dietary supplements (Airborne) will protect them from the disease (a cold). Despite easily available information busting the myths of plane travel and the effectiveness of Airborne, and what we actually know about cold viruses and catching colds, this myth set remains present and powerful. If such a small myth set that is relatively inconsequential to our daily lives can be so pervasive in the face of available evidence against it, what does that say about much larger, more ingrained myths about human nature?

Good question. To move forward into our consideration of major myths in human nature and why we should be really concerned with them, we first need to think a bit about what we mean by "human nature" and why it is important to the way we construct ideas about ourselves in society.

What is human nature for most people?

All studies of man, from history to linguistics and psychology, are faced with the question of whether, in the last instance, we are the product of all kinds of external factors, or if, in spite of our differences, we have something we could call a common human nature, by which we can recognize each other as human beings.

—Fons Elders, philosopher[8]

So that in the nature of man, we find three principal causes of quarrel. First, competition; secondly, diffidence; thirdly, glory. The first maketh men invade for gain; the second, for safety; and the third, for reputation. The first use violence, to make themselves masters of other men's persons, wives, children, and cattle; the second, to defend them; the third, for trifles, as a word, a smile, a different opinion, and any other sign of undervalue, either direct in their persons or by reflection in their kindred, their friends, their nation, their profession, or their name.

—Thomas Hobbes, philosopher[9]

It is true that I mistrust the notion of human nature a little. . . . In the history of knowledge, the notion of human nature seems to me mainly to have played the role of an epistemological indicator to designate certain types of discourse in relation to or in opposition to theology or biology or history.

—Michel Foucault, philosopher[10]

Is there a human nature? Is it based on individuality, competition, and the struggle for survival? Or is the whole concept of human nature just a tool to help us think about differences between ways of asking questions about being human? The usual definition of human nature as the general characteristics, feelings, and traits of people does not really tell us much. Philosophers have spent several millennia debating the subject, but generally seem to answer it by questioning the ubiquity of any human nature or proposing that human nature is animalistic and brutish. Many philosophers, especially those emerging from the Enlightenment, such as Jean-Jacques Rousseau and John Locke, suggest that it is in our nature to move past the animalism at our core and toward a set of agreements by human beings to surrender our individual rights in order to secure the protection and stability of a social organization (eventually a government).

However, most people do not delve into the writings of philosophers to understand human nature. Today, because of our increasing knowledge and expectations about genetics and biology, we frequently turn to scientists, largely biologists and psychologists, for explanations of what a human nature might be. In many cases these scientists agree that there is some core to humanity but disagree as to how we should best envision it as interacting with the world around us, especially with that pesky thing we call culture.

Here are two quotes from different scientists that seem to express opposing opinions about human nature:

> Human behavior is a product both of our innate human nature and of our individual experience and environment. . . . We emphasize biological influences on human behavior, because most social scientists explain human behavior as if evolution stops at the neck and as if our behavior is a product almost entirely of environment and socialization. In contrast, evolutionary psychologists see human nature as a collection of psychological adaptations that often operate beneath conscious thinking to solve problems of survival and reproduction by predisposing us to think or feel in certain ways. Our preference for sweets and fats is an evolved psychological mechanism. We do not consciously choose to like sweets and fats; they just taste good to us. The implications of some of the ideas in this article may seem immoral. . . . We state them because they are true, supported by documented scientific evidence. Like it or not, human nature is simply not politically correct. (Alan Miller and Satoshi Kanazawa, evolutionary psychologists)[11]

> . . . there is no such thing as a fixed human nature, but rather an interaction between our genotypes, the genetic information we have and the different environments we live in, with the result that all our natures are unique. . . . [I] emphasize the gigantic role that cultural evolution plays in making individuals different, and in making groups different. I'm hoping to counter a view that I'm afraid is all too common among the American public, that all of our behavior is controlled by our genes, and that there are genes that code for aggressiveness, inquisitiveness and so on. The truth is, you can never remove culture from the mix. (Paul Ehrlich, biologist)[12]

Can both of these quotes be right? More worrisome, if scientists don't agree on human nature, what about the rest of us? Perhaps song lyrics and Wikipedia might be more reflective of how people every day get some idea of what human nature is. These common resources can tell us that there is a human nature and an inherent set of characteristics that are unavoidably part of being human. The popular take on the theme suggests that going against this human nature is difficult and that if we express our basic selves, our inner drives, even in the face of negative reactions from society, what we get is a reflection of our human nature:

> Human nature refers to the distinguishing characteristics, including ways of thinking, feeling and acting, that humans tend to have naturally. (Wikipedia)[13]

> You cannot go against nature
> because when you do go against nature it's part of nature, too
> (Love and Rockets, band)[14]

> Express yourself, don't repress yourself,
> And I'm not sorry,
> It's human nature
> (Madonna, singer)[15]

Certainly most people (philosophers, scientists, pop stars, and the public) would agree that there are many things that unite humanity, but whether there is a specific set of behaviors and proclivities that could be called a "nature" is not quite so clear. To understand why we are reviewing what "human nature" might mean, let's think about this term and the power it has in our society.

We know that, at a genetic level, there is indeed a unity at the base of humanity. We all share the same basic set of DNA (more detail on this in chapter 3), are the same species, and thus share a common general biological history and a set of the same distant ancestors. This is a fact. However, does this mean that there is also some specific set of behaviors or drives and desires that emerge from this shared DNA and biological histories? This is what those who argue for a specific human nature would suggest. This proposition is a very attractive one and many folks buy into it (at least a bit) and also see men and women as having slightly different specific sets of behavior stemming from this genetic and evolutionary origin (that is, slightly different natures). Although many would not admit it, there is probably also a significant set of folks who would accept that the different races also have some differences at this basic, "natural," level. A common popular perspective is that we all share a general human nature, which we can see expressed as behavior and proclivities, and that different divisions of humanity (sexes, races) might have their own specific patterns emerging out of this nature.

Why is this important? If this view of human nature is popular, we encounter a problem of conflating "natural" and "right" (or "is" and "ought") with the misperception that we can identify one true way to be human. This stance has been much debated and discussed by philosophers and scientists for centuries, and has been an especially hot topic over the last five decades. If there is a specific natural way to be a human, then one could rightfully argue that this is the way humans evolved (or were designed) to be. It then follows that if we have societal rules and expectations that contradict or inhibit these natural drives and behaviors, we might see problems as humans try to conform to societal rules and "go against" their natures. This view assumes that the behaviors and drives emerging from our nature are more correct, and set deeper, than behaviors emerging from externally influenced sources (like societal expectations). In other words, this is a position that can explain and accept human behavior, regardless of its implications, by invoking our inner nature as a justification. These are exactly the kind of societal myths that we are tackling in this book.

Let's think about a few examples of how this might play out. The most prominent is the idea about a biological core and a cultural coating—that we have a basic nature and that our culture conceals our more savage and animalistic side. One of the best depictions of this idea comes from the *Lord of the Flies,* the famous novella by William Golding. In his story a group of young and well-behaved English schoolboys are shipwrecked on an island. As time wears on without rescue the boys begin to shed the restraints of society; their cultural veneer is slowly stripped away. First, they begin to change their behavior and dress, chanting and dancing around the fire, painting themselves, and wearing less and less clothing. They become savages, reflecting our more primitive (natural) selves. The stronger boys begin to dominate the weak through social intimidation and physical threat. Only one boy seems to hold on to his humanity (the protagonist), trying in vain to maintain a just and civil society. But in the end, even his basic instincts of violence and savagery emerge. It's a commonly held belief that if you strip away culture, that which keeps us well behaved, then a beastly savage will emerge (especially in men).

This kind of idea about what a person's basic self is can be played off what the individual (or society) tries to make of them. If the person fails, goes astray, or deviates from societal expectations we often say "that is just the way he (she) was born," implying that despite the trappings of society, for some people (many people?) their nature is constantly trying to push through. This can be utilized to create and justify myths about human nature, or even to alter criminal cases.[16]

Interestingly, this way of thinking about human nature has both religious and atheistic analogues. That is, it does not matter whether you believe our genetic and biological heritage set the baseline for what we do and why we do it or whether you believe that a deity created humans and set the baseline from which we occasionally stray. For example, one of the foremost atheists and proponents of a strong genetic basis for humanity's inner core, the zoologist Richard Dawkins, argues that a human nature is created by the competition between selfish genes resulting in selfish behavior in regards to food, health, and reproductive success for the humans whose bodies the genes are in. Therefore, our evolutionary history sets up why we do what we do. However, he also notes a major role for culture and even a sort of free will wherein humans can be cognizant of their nature and thus make active decisions to try and go against it.[17] On the other hand Thomas Aquinas, arguably one of the most influential philosophers and Christian theologians of the last millennia and the propagator of the philosophy of natural law, tells us that humans have basic instincts that they share with all other animals. He suggests that these instincts emerge as desires (often selfish desires of food, drink, sexual activity, etc.) but that reason acts to mediate the expression of these desires and that humans act together (via reason) to create rules that enable us to satisfy the core of the desires in moral and useful ways (society). However, for Aquinas this essential human nature (part of what is called natural law) and our unique human ability to reason are all creations of God, not our biological histories. It is the ability to reason that enables us to escape the basic, animalistic facets of our inner nature (a form of free will, enabled or created and set in motion by a deity).[18] Whether we attribute our inner selves to a deity or to genes (DNA), the result is the same. There is an assumption of a competitive, animalistic drive that might be reined in by human rationality, society, and our cultural actions.

It is important to emphasize that I am not leading toward the argument that there is not a shared evolutionary heritage, biological similarities, and shared patterns of behavior across our species; there is. Those who would argue that we are a blank slate at birth, that our social and experiential histories are all that count in making us who we are, are wrong. We examine this in the next two chapters and then later when we tackle the major myths themselves. I am not making a purely "nurture" argument for human behavior. Much of what we have covered so far, and much of what is prominent in both the popular and the many academic views of this topic is from the nature-nurture debate.

Nearly everyone agrees that both biology and culture are important in human behavior, but many want to be able to parse out how much of our genes or our culture is responsible for any given human action or set of beliefs. However, as we will see from busting the myths about race, aggression, and sex differences, it is almost never an either/or situation with nature and nurture. In fact, human behavior is almost always "naturenurtural";[19] it is a true synthesis and fusion of nature and nurture, not just the product of adding nurture to nature. There are not two halves to being human. When we think about humans it is a mistake to think that our biology exists without our cultural experience and that our cultural selves are not constantly entangled with our biology. Simple examples of this kind of engagement can be found in things like our adult height (a product of the integration of at least genes, diet, physical experience, climate, and conditions of health and disease) or the ability to throw a baseball and kick a soccer ball (which can be influenced by integration of at least health, height, training, structure of lower limb muscles, altitude, nationality, sex, gender, peer group, and diet). We go into this in more detail in subsequent chapters, but be very clear that humans are neither a blank slate nor preordained entities; both of those perspectives miss the boat: we are naturenurtural.

SCIENCE, SOCIETY, AND IGNORANCE: IGNORANCE IS NOT BLISS, IT'S JUST IGNORANCE

What is wrong with the statement "ignorance is bliss"? Ignorance usually means a lack of knowledge or information. The idea that not knowing something can make a person happy is a fairly common one. This is usually associated with children being happy because they might not understand the concept of death, the flipside being that adults are unhappy or burdened with the knowledge of their eventual death. This is encapsulated in the oft-quoted line "ignorance is bliss" from the British poet Thomas Gray's poem, "Ode on a Distant Prospect of Eton College" (1742). It is the last line from which we get the saying: "Where ignorance is bliss, 'tis folly to be wise." But it is worth reading at least the entire last stanza to understand the point:

> To each his Sufferings: all are men,
> Condemned alike to groan;
> The tender for another's pain,
> The unfeeling for his own.

Yet ah! why should they know their fate?
Since sorrow never comes too late,
And happiness too swiftly flies.
Thought would destroy their paradise.
No more; where ignorance is bliss,
'Tis folly to be wise.[20]

Written shortly after his graduation from Eton, the poem seems to be an ode to the happy times as a student and the potentially terrifying realities (knowledge, life, responsibilities) that await the graduate. Many of us hold some idealistic notions that we are happier at times when we do not really know the big picture and are able to live in our clamshells, insulated from the negatives of the larger world. As with Gray, this is often the case for those of us lucky enough to have gentle and protected upbringings, involving good schools and supportive environments. The concept that thought (knowledge) destroys paradise, that not knowing certain things is a blessing, can be a nice literary turn of phrase, but it is a dangerous tool when applied to ideas about human nature.

Throughout the main portion of this book, as we bust three major myths of human nature, we are going to be combating a widespread ignorance of what we actually know about human beings. Ignorance—the lack of knowledge or information about human behavior, biology, and history—only acts to inhibit our ability to understand who we are and why we do what we do.

The concept that there can be too much knowledge is ridiculous in this context; we need to have access to as much information as possible to be able to make up our own minds. We can never know everything, nor even nearly as much as we might like. However, we should all have access to the information and the basic skills with which to assemble and critically assess the information. In this case, ignorance is in part the result of active concealment on a few issues, but mainly the result of a real lack of science education and a divide between academic scientific knowledge and the public which needs to have access to this scientific understanding. The information we need in order not to be ignorant about human nature lies in many different places, mostly in academic contexts separated from the public and even from different types of researchers.

Okay, so how does not knowing much about science and biological and historical details about humans put us at a disadvantage when trying to think about human nature? First of all, we may be much more easily manipulated. If we do not know what information is actually available or how to use specific information to begin to ask effective questions, then we might not even recognize relevant information when we come across it. If we do not recognize specific types of a knowledge as relevant, or know how to apply that information, our ability to use that information to affect our lives and surroundings via active choice is reduced, and even simple decisions might be constrained.

Take food labels, which are a perfect example of the need to integrate basic scientific knowledge and general public information. When shopping the aisles of an American grocery store we may notice the panel of nutritional facts on food labels. On this panel is a description of the contents, energetic values, and major dietary components in the packaged food. This information is meant to allow us to make an informed choice about what we eat, if we wish to protect our health and well-being. The panel begins with the amount of calories and calories from fat, total fat, different types of fat (saturated, trans, polyunsaturated, etc.), amounts of cholesterol, sodium, carbohydrates, fibers, proteins, and a variety of vitamins. Often, at the bottom of this panel there is also a chart showing how much of each item should be consumed, depending on the daily calories in one's diet (usually a choice between either 2,000 or 2,500). Also on the package label is a list of all the ingredients found in that foodstuff.

What do we need to know to be able to use the scientific information on this label to our benefit? First of all, we must have a basic understanding of math and the measurement system of our society. The items are generally listed on a per-serving size given in both English and metric units (such as ½ cup, 3 oz., 2 tablespoons, 50 ml, 40 g, etc.). To calculate the actual amounts requires that we know how much a teaspoon, a cup, an ounce, a milliliter, or a gram is. This is one area where there is some manipulation on the part of food packagers. By listing fairly small and unrealistic serving sizes, they create an artificial picture of what is contained in the actual serving that one eats. For example, some soft drink cans list one 12-ounce can as two servings, but how many people typically drink half a can of soda and save the rest for later?

After math, we need to know what a calorie is (it is a unit of energy) and why it matters what percentage of calories are from fat. We need to know what fat is (oils and lipids, which are organic chemical compounds of glycerols and fatty acids) and what that has to do with our body. Then we need to know about cholesterol (a steroid metabolite) and sodium (salt), and finally carbohydrates (sugars made of carbon, hydrogen, and oxygen molecules) and fiber (the indigestible portions of plants) and the interactions between them. And we have not even broached the subject of actual ingredients. This is a lot of very detailed and complex knowledge about society, biology, and mathematics. How do we usually deal with this? Ignorance and myth.

Many people now realize that we should be aware of what the food labels say. However, most of us use a mixture of commonly held assumptions (societal myths) and a little bit of knowledge about health to read the labels. Many of us gloss over the labels and make a calculation based on the assumption that lower fat and/or lower calories is good. Or, we look at the carbohydrates, and make the assessment that more is bad and less is good. Others might know that higher fiber alters what carbohydrates do in the digestive tract so they do a quick calculation of the fiber-to-carbohydrate ratio. Do most people know what a fat molecule is or how many calories they both burn and consume in a day? Do most of us even know what that means? Do we know what "good" and "bad" cholesterol are and how that relates to the cholesterol label on foods? No, in fact many people rely on a bit of knowledge and a lot of popular assumptions about what these terms mean. The main point here is that we actually have a good deal of information available to us (the data are there on the food labels), but the vast majority of us are ignorant of what the information means and/or how to use it to make a good assessment of what is best for our dietary needs and health. So, we wing it.

But in this case, winging it is really taking what we hear on the television, read on the Internet or in books, and get from friends, as our baseline knowledge. A good deal of this information comes first, second, or third hand from so-called experts, scientists and medical doctors who are pitching an item, writing a book, blog, or Web site about health, or are featured on a television program. As a society, we have a tendency to believe what we think comes from the realm of science without fully understanding what science is or how we should approach information delivered by scientists.[21]

Should we listen to what scientists have to say?

Yes and no. For example, James Watson, the Nobel-Prize-winning geneticist, former director of the Cold Springs Harbor Research Institute, and codiscoverer of the structure of DNA, told a British audience in 2007 that he was "inherently gloomy about the prospect of Africa" because "all our social policies are based on the fact that their intelligence is the same as ours—whereas all the testing says not really."[22] The knowledge that Watson helped produce regarding the structure of DNA remains one of the most well-tested and well-supported important contributions to modern genetics. However, his thoughts on race, Africa, and intelligence are opinions that contradict the well-tested scientific knowledge available about how race, geography, and genetics in humans actually work (see chapter 4). Here is a case where the scientific knowledge is available, but the scientist is stating a personal opinion regardless of the available information. Conflating the thoughts and opinions of scientists with science itself is a source of much of the misinformation leading to myths about human nature.

We should care about the knowledge generated by scientific projects, but not always so much what scientists themselves have to say. The word "science" is a powerful one in our society and popular explanations associated with science and scientists have a lot of weight in our views of what humans do. However, we need to be careful about what we actually mean by "science" and pay close attention to how and where scientific knowledge is produced and disseminated. For example, the anthropologist Jon Marks critiques the notion that science reflects a neutral investigation of reality but at the same time contends that it can produce real and important results:

> Science is the production of convincing knowledge in modern society. . . . By using production we acknowledge that science is not a passive experience. Scientific knowledge is a product—and as a product it is the result of some process. . . . There is a subtler and more threatening point embedded in this recognition, however. If science is the active production of something—say, reliable information about the universe—then it is more than, or at least different from, mere discovery. Discovery is a passive operation: to a suitably primed observer, the fact merely reveals itself . . . the production of scientific knowledge is highly context-specific, and . . . it is the context, more than the particulars of the discovery, that are critical. (Jonathan Marks, anthropologist)[23]

Marks, as many have done before him, focuses on how scientists (the people doing science) affect the kinds of questions asked, methods used, and results obtained. He notes that science is a process that can lead to discovery, but the specific context from which the discovery emerges can be as important as the facts themselves. Take the example of the discovery that the human immunodeficiency virus (HIV) can lead to acquired immune deficiency syndrome (AIDS), for which there is treatment but no cure. We have an enormous amount of scientific knowledge about HIV and AIDS, but at the same time there are numerous scientists and other figures offering opinions about the origin of the virus and disease that are not based on the scientific data available.[24]

Much of the knowledge produced via the method we call science is valuable to us because it can be tested and verified. In reality, science is not a thing or a system of thinking, it is a method. That method is more or less as follows: observe an event or occurrence, create a testable explanation (a hypothesis) for the event, and test it. If the test is refuted, showing the explanation to be wrong, develop another testable explanation and test that. If the first test is positive, then retest it numerous times and maybe have other people run the same test to see if they get the same answer. If all tests support the explanation then it is a viable explanation (hypothesis). If a bunch of supported hypotheses can be put together to create a broader and more robust general explanation, then that becomes what we call a theory. Now in practice today much science is taking other people's observations, or things that we know occur, and creating lots of different hypotheses and testing them. Many people practicing science no longer collect the initial observations themselves but rather, use the gigantic body of available knowledge to create more refined and detailed explanations that they can then test.

Gravity is a perfect example of this. If we drop something from a sufficiently high spot anywhere on the planet it will eventually accelerate to about $9.81 m/s^2$ (or 32.2 feet/s^2). This is a fact discovered through observation and scientific inquiry. Scientists explain the reason objects fall at that rate by the theory of gravity. Gravity is not a fact. It is a well-tested explanation of the fact of things falling toward earth when we drop them. Thousands of scientists over the last three centuries have questioned and tested the theory of gravity and found it supported. This is science; it is a process of examining the world around us and reducing the number of possible explanations to a few probable explanations via testing.

When we talk about facts in science, we are talking about something very different from facts in a court case. In a scientific context, a fact has to be the same for everyone (i.e., when falling, all objects accelerate at the same rate). In the courtroom, in journalism, and in popular usage a fact is more along the lines of something that appears to be indisputably accurate, or more generally, facts are simply information used as evidence or to support a position (regardless of their actual accuracy). This is where the popular understanding of science, facts, and scientists becomes complicated. Scientific facts are few and far between, popular facts are commonplace and not usually facts at all.

This is an important point because in our society we often conflate the science and the scientists. Or even more dangerous: we conflate any specialist in a science-related field with the facts from that field. For example, many advertisements use the tagline "Four out of five dentists prefer product X." You can substitute doctors or other professional specialists for dentists and the result is the same. Note that the ad does not say "Data collected and hypotheses tested by the following three reputable research laboratories show that . . .," but rather it relies on the opinion of specialists, not on science. Now, one might argue that the opinion of specialists is informed by science, and that is possible, but it remains clear that science itself does not receive payments, create advertisements, speak publicly, publish books, or issue press statements.

Our understanding of scientific assertions and knowledge is very important because it is considered specialist knowledge, and our society deems specialist knowledge, especially if it appears to be rooted in biology or something related to science, as more relevant to discussions about human nature than other types of knowledge. The anthropologist Jon Marks credits the physicist and novelist C. P. Snow with forcefully acknowledging that we can see the field of science as an anthropological culture and that it is, at least in part, culturally constructed. This is particularly relevant in the context of this book. I intend to demonstrate that a major portion of the three myths of human nature that we will bust are based on culturally constructed concepts that incorporate some science but a greater amount of opinion (including the opinions of scientists) that is not supported by the actual knowledge derived from the science itself.

Think back to the two examples of popular myths we have discussed so far in this chapter: the link between airplanes, vitamin C, and colds, and the reluctance of men to ask for directions. Then think of what we briefly reviewed regarding what we know, from testing and analyses, about plane travel and air circulation, cold viruses and their impact, male and female similarities and differences, and human evolutionary history. Now think about people's opinions on these topics and where those opinions might be coming from. The power of specialist opinion (and advertising) is particularly strong in our society. Think about the assertions for the dietary supplement Airborne and the claims of books about understanding and negotiating male and female differences. There is a tendency to appeal to some biological core, to get at the base of being human and, nearly always, that base is considered more natural if it is tied in some way (even superficially) to our biology. This brings us back to the nature-nurture fallacy. This is one of our biggest hurdles to busting major myths in human nature. As humans we are not one or the other, we are always naturenurtural. The problem with this perspective is that to tackle it, to use such a complicated viewpoint, we need to draw and integrate information from a wide variety of sources, something that is increasingly difficult in our society.

Why should we try to integrate different types of knowledge in our attempts at understanding?

The ability to really get to the heart of topics that deal with human behavior requires us to synthesize bits of information from many different sources. Even to understand the simple examples of the success of Airborne and the humor in the joke about men not asking for directions we need a bit of chemistry and engineering, some ideas about biology and health, some familiarity with human physiology and sex differences, and knowledge about the history of gender perspectives in our society. Unfortunately, our education system does not always do a good job of getting us the skills to integrate different types of information. In math classes we do math, in history class, history, and in biology class, biology. This is important when we are trying to get basic skills down, but as you get older and experience more and more in life you need to be able to integrate information from different subject areas to really get good pictures of what is going on.

Let's take the example of women's voting rights. The Nineteenth Amendment to the US Constitution passed in August 1920 gave women the right to vote. Up until that time, women in the United States were not allowed to cast ballots for public office (although a few western states gave women some voting rights in the first decade of the 1900s). Most people would think that to understand why this was the case we need information from history and maybe political science, and we do. We need to know about the Constitution, the folks who wrote it, and how it expanded from the 1780s to today. We also need to know that the women's suffrage movement kicked into high gear in the mid to late 1800s and was tied in with—and often opposed to because of this tie—the right to vote for American males of African descent (the Fifteenth Amendment, which passed in 1869). The Fifteenth Amendment expanded who could vote but still defined a "citizen" as male (only then as both white and black males) and thus was actually opposed by many in the women's suffrage campaign. However, in addition to this cultural and political history we need to know a bit of biology, psychology, and to think about popular myths. It turns out that one of the main arguments against giving women the right to vote was based on a popular myth about male and female differences. There were serious expectations about women's roles in society and their need to be in specific social positions (homemaking and caretaking, for example) relative to men that were challenged by an expansion in women's roles both during the Civil War and World War I.

As women displayed more public abilities to participate in a wider array of social positions, many men looked for deeper justifications for keeping them out of politics. This justification found its foothold in the form of "natural" differences between men and women. As famously noted by the Reverend O.B. Frothingham, in a essay in *Arena Magazine* in 1900, women had a natural preponderance of "feeling" relative to men, which disqualified them from acting in the sphere of practical politics. What Frothingham was really saying was that women's "natural" inability to be practical and logical made them unsuitable for participating in the complicated and nuanced political reality of voting and running the government; they made decisions based on feeling, were sweeping and overly general, and never nuanced.[25] This is a myth that had great impact then (and remains somewhat alive today). To understand the validity of the Nineteenth Amendment we must understand why this myth is incorrect. We have to know a bit about human biology and behavior, including that male and female brains are not different in their ability to perform logical analyses and that our limbic systems (that are the basis of our emotions) are largely identical. A bit of psychology (to understand why people thought about these differences this way) and a bit of cultural anthropology and historical archeology to tell us about how men and women really lived in that time period (a lot more overlap than we are commonly led to believe today) are core to our ability to get a full picture of what was going on and why it was happening.[26]

Another good example of the need to mix different types of information comes from the hypothesis about a link between slavery and salt. First proposed in the late 1980s to explain higher rates of hypertension (high blood pressure) in Americans of African descent, this idea was rapidly picked up by the popular press and gained credence due to a number of myths in our society. In a nutshell, this hypothesis suggests that blacks are genetically biased toward having high blood pressure because peoples from West Africa (who were transported to the Americas during in the slave trade) were adapted to low-salt environments; the forced voyage across the Atlantic on slave ships and the harsh life on plantations favored those individuals with high ability to retain salt (giving them better resistance to diseases that caused diarrhea and dehydration). The presumed genetic proclivity to retain salt was then passed to descendants of slaves, leading to the problems of hypertension in the high-salt environments of today. Fascinating story, but not true.[27] The slavery and salt story rests on the myth of races exhibiting real biological differences. To realize that this is a false set of ideas and misrepresentation of data we need to know about the ecology of West Africa and the plantations in the South and Caribbean, the slave trade and the transatlantic journeys, the differences between slave conditions and health in the United States and Caribbean countries, the salt trade (that occurred alongside the slave trade), the reality of hypertension in the United States across all ethnic groups, economic classes, and regions, and a bit about human biology and genetics. Getting information from archeology, cultural anthropology, physiology, genetics, history, and evolutionary theory is key to busting the myth of slavery and salt as an explanation for a "naturalness" in health inequalities in the United States.

My point here is that the lack of an effective integration of biological, anthropological, and evolutionary knowledge (at a minimum) with societal perspectives and popular discussions can dramatically inhibit our understanding of our histories, our daily lives, and of what human nature might be. Without these kinds of integration we remain ignorant about many core areas of being human and human histories and are susceptible to participating in, and propagating, false myths about human nature. The need to integrate across different areas of knowledge also comes with a suite of problems. Where do we find the information? How do we select the right information from the vast amount out there? How do we understand details in fields that we know little or nothing about? All good questions and in the appendix of this book we tackle a few of them. However, the goal for the bulk of this book is to distill the appropriate information from diverse sources into nuggets that we can use and place them in a context that makes sense to the general reader. To start this distillation, chapters 2 and 3 focus on the interrelations of human development, the realities of culture, and evolution and biology. Then in chapters 4, 5, and 6 we use this contextual perspective to assess data available for busting three major myths about human nature.

Culture—Problems with What We Believe about Being Human

My dissertation advisor at Berkeley, Phyllis Dolhinow, once began a lecture about research and the interpretation of information with the phrase "I would not have seen it if I hadn't believed it."[1] This twist on the adage "seeing is believing" pinpoints an aspect of being human that makes myths about human nature so resilient. What we believe to be true affects the way we see the world, the way we think things should be, and how we interpret information presented to us. That our perspectives affect our perceptions on a daily basis is a core concept of this book.

Professor Dolhinow used this reversal of the adage to describe how many researchers become steeped in specific perspectives about how primates (including humans) behave.[2] Through reading only a selection of the available published reports and being trained by individuals strongly committed to a particular way of thinking about evolution and ecology and how it shapes primate behavior, researchers develop a set of expectations about what they are going into the field to observe. This set of expectations can influence the way they record and conceptualize what they see. For example, in 1999 in Bali, Indonesia, a primatologist colleague and I were watching a group of macaque monkeys when a series of interactions between a few adults and an infant took place. Standing side by side and having seen the same series of behaviors, we turned to one another and excitedly described two slightly different things. My training and perspective influenced me to focus on the adults and their behavior toward one another, not considering the infant as a main player in the interactions. My colleague, trained in a different perspective, looked specifically at the actions of the adult males as they related to the infant, not focusing as much on the adult females. We each had a set of expectations about how things are that colored where we looked and how we described what we saw. We were exposed to the same set of information but perceived two different outcomes. Once I realized what had happened the experience stuck with me: we saw the same actions, but our beliefs about the world led us to actually experience different things.

Humans are extremely biased in how they interpret information regardless of whether it comes by sight, sound, smell, taste, touch, or some mixture of these senses. Our physiological abilities (hearing, sight, olfaction, etc.) vary across people, but it is primarily our cognitive interpretations of these senses that act as the filter.[3] The route by which we develop this cognitive filter is the process of learning how to be a member of a society, ongoing our whole lives. Social scientists call this enculturation and development, frequently dividing the acquisition of culture from the development of the body. However, as noted in chapter 1, such a marked separation misses the true complexities, the biocultural reality, of being human.[4] Humans are almost always naturenurtural, and in order to think about the myths of human nature that we are targeting in this book we need a baseline of understanding about how we come to be who we are. In chapters 2 and 3, first we will examine in greater detail how humans develop and acquire specific perspectives about the world around them. Then we will temporarily disentangle culture and biology to discuss what we know about how each works, and finally we will get back to the messy reality of being human and provide the basic tool kit necessary for delving into the information in chapters 4–6 and doing a bit of myth busting.

WE ARE WHO WE MEET

In teaching introductory classes in anthropology, I use the following story as an illustration of what it takes to become human. A long time ago there was a king who had an ongoing disagreement with his chancellors and clerics. He was certain that English was the original language, while others in his court argued for Latin, Hebrew, or Aramaic. The discussion went round and round without any resolution. Finally, the king had an idea: he would discover what the original language was by keeping five newborn babies in silent isolation, never hearing a spoken word. Thus, when they finally spoke, in whatever language the babies uttered, their first words must be the original language of man. The king arranged to have five newborns taken from their mothers and set up in five towers isolated from towns, villages, and people. Each infant had a mute wet nurse who would feed them a few times a day and have no other interaction with them at all. The infants would have no interactions with any other people at all until they spoke, keeping them pristine for the experiment.

At this point in the story, aside from a few students who protest the inhumanity of the king's experiment, most students ask, "Well, what language, if any, did the babies speak?" To which I respond, "None, they all died." The point is that if you isolate a human infant from human contact and society, you do not discover the innate and original nature of being human, you will just kill the infant. Or if by some miracle the child lives, you will have a very damaged individual—biologically, psychologically, and socially. While I am pretty certain that this story about the king and his babies is not true, it does illustrate the common misconception that there is an innate core to humans that is altered, masked, or otherwise concealed by culture. This is the wrong way to envision the scenario of becoming human: humans need to be around each other for social, physiological, and psychological reasons, and becoming (and being) human is a process that is simultaneously biological and cultural. We need to grow up around one another to be fully human.

What we think of as "normal," what we consider intuitive knowledge and common sense, rarely emerges from some inner biological core subconsciously telling us what is "true." Rather it is more likely to be the result of the experiences we have had throughout the course of our lives and the way in which these events interact with and shape or influence our bodies and minds. As early as 1930 one of the core figures in American anthropology, Franz Boas, noted this and saw that we are influenced by the world around us and that, simultaneously, our actions shape and influence that world as well. Although we are biological organisms, the totality of the human experience cannot be reduced to either specific innate (biological) or external (environmental) influences; it is a synthesis of both—we are naturenurtural. The anthropologist Tim Ingold tells us that "to exist as a sentient being, people must already be situated in a certain environment and committed to the relationships this entails," and these relationships are built up and modified over the course of our lives.[5]

This chapter presents an overview of these relationships by examining enculturation, development, and how this knowledge shapes who we are and our vision of humanity and human nature.

Developing Schemata as Well as Bodies

Each human has a unique genetic complement: our bodies develop in particular ways that depend on our mother's health and her life while she is pregnant with us, on the nutritional and physical context in which we are born, and a myriad of factors that shape our bodies and minds as we develop. Because we all grow up in a particular society we develop a particular way of interpreting the world. This development is influenced by those around us, our cultural history, and the ongoing patterns in our society, our community, and our family. Through these influences we acquire a significant portion of our *schemata*—or in nontechnical terms, our all-encompassing worldview, the way we see and interpret the world around us.[6] Many researchers think of this as enculturation (literally "getting cultured"), mainly in terms of social attributes. However, recently there is a trend to think more broadly about enculturation as a process of *enskillment*.

The term "enskillment" encompasses the way humans learn enormous amounts by imitation (conscious and unconscious) and absorb cultural knowledge from the communities they reside in or experience. Tim Ingold suggests that our intuitive understanding of the world, and the ways in which we use our bodies in it, comes from "perceptual skills that emerge, for each and every being, through a process of development in a historically specific environment."[7] These perceptual skills mesh with the physical and cognitive aspects of our development to help shape how we act and perceive ourselves and our world. For ease of general explanation we can divide this process of enskillment into three areas: physical development, bio-enculturation, and social context.

Our physical development shares a great deal in common with that of other mammals, and more specifically other primates. However, there are also a few very significant differences. In general, we share the common mammalian pattern of birthing babies that need a good deal of assistance from the mother. Specifically, we are like many other primates (especially apes) in that we are born with large brains but dependent on our mothers and other group members for physical and social care for an extended period. However, humans are born more underdeveloped physically than any other primate. We cannot move on our own, feed ourselves, or even communicate effectively as infants. To survive their first decade, humans are even more reliant on mothers and others than any other species.[8]

Because our brains are so large (relative to our bodies and the brains of most other animals), they take a long time to mature. This means we spend most of our youth training the brain to work (providing us with both neurological connections and functions, and a set of perceptual skills). This slow maturation makes us sponges for information during our youth. A byproduct of our slow maturation is a pretty long life span during which we are constantly seeing, smelling, touching, and otherwise experiencing the world around us. This is translated into new neurological connections and physiological patterns. At puberty a series of physical changes in our bodies interfaces with social contexts (our cultures' expectations and gender roles) to produce what we term adults. As we age, further physical changes (menopause, body shape, bone density, for example) occur. Throughout our physical lifespan our bodies and minds interact via patterns that are influenced by our specific environments. What our body and mind perceive as normal is highly contingent on where and how we mature.

A simple example of this is our body's response to heat or cold. If you grow up in a very hot environment when you are an adult your body will respond more effectively, and less noticeably, to heat stress and you will perceive heat in very different ways than someone from a largely cold environment.[9] The hot-as-normal person would sweat less and feel comfortable at higher temperatures than the cold-as-normal person. So these two could be in the same place at the same time in the same temperature, but their bodies would respond quite differently and they would perceive the local environment differently (even if they were twins separated at birth). Cognitively, both these two individuals have accurate perceptions of the heat, and each will feel that their interpretation and physical response are natural, but physiologically each will respond quite distinctively as each has a slightly different history and skill set for interacting with the environment.

Another simple example of this is the foods one grows up eating. The adage "you are what you eat" is better understood as your body and mind creating schemata and physical responses to food depending on what was put into your mouth from birth. Growing up eating primarily hot peppers, root crops, and other vegetables is going to produce an adult with a very different psychological and physical sense of food than one who grew up eating primarily whale fat and fish. This is in spite of the fact that the biological mechanisms of taste and smell vary little across human populations; rather, the social and physiological skill set, our perceptions, are highly dependent on experience. Two individuals can taste a single food, new to both of them, and give different reports on its taste and smell, even though, chemically, their tongues, noses, and brains are receiving the same stimulus.

The same-food, different-taste example could also be seen as bio-enculturation, but generally we conceive of this in a social context. While everyone has heard the saying "you are what you eat" few realize that when it comes to enculturation the saying "you are who you meet" is also very accurate. Those individuals we encounter on a daily basis, speak with, learn from, and hear about regularly are core to our developing perceptions. Our social development, schooling, gender acquisition, peer group interactions, and parental and sibling interactions have an enormous impact on shaping our schemata and how our brains respond to social stimuli. These patterns that we participate in and that surround us daily help shape our perceptions of what behavior, language, and mannerisms are normal and natural in our world. Not only do we learn from observation and direct instruction what is considered normal behavior, but the body's perception systems also have a series of built-in mechanisms for taking what we are regularly exposed to and making it a part of our neurological makeup.

For example, recent research into neural systems of the brain demonstrates a series of areas called mirror neurons in monkeys and humans. These areas facilitate our understanding of what someone else is doing (or possibly feeling) by mapping what we see onto an internal, neurological simulation of the same movement. In other words, a person can interpret and imitate the actions of another person they are watching simply by simulating the other's behavior in their own brain via the action of a set of mirror neurons.[10] This suggests that what we see around us, the way people walk, talk, wave, sit, dance, hug (or not), and so on can be translated into specific neurological pathways that become a part of us. Humans (and maybe monkeys and apes) are wired to pick up social cues and behaviors and integrate them into their basic brain functioning. We know that the way movement and behavior occur with humans, while constrained by basic physical parameters, are widely variable across cultural contexts. So, bio-enculturation via mirror neurons and a wide range of physiological exposures to social and environmental contexts also contribute to building and maintaining our schemata.

For example, take something as simple as personal space. In some societies people engage in a lot of physical contact when greeting (hugging, kissing, etc.), even with strangers, and in other societies there is very little contact (maybe a handshake). If you grow up in the little-contact society and then are immersed in the high-contact society, you might be socially uncomfortable, and you will also probably be physically uncomfortable. Your mind and body have been bio-enculturated in a particular way regarding space and touch so that when in a situation where these expectations are disrupted, you respond both cognitively and physically.

In addition to our physical development and bio-enculturation, the people and society around us set up the types and contents of inputs into our developing schemata. The basic expectations of a society, the social norms and rules, are a sort of baseline for what we incorporate into our perceptions of the world. The history of a society and the ideals and culturally salient events that are highlighted in it create a shared set of social expectations for its members.[11] In the United States we tend to highlight concepts of individual liberties and freedom of expression, elements that are heralded as natural rights for all humans. This gives us a piece of our cultural schemata about what is inherently true about a human's rights.[12] The social context is not a fixed entity, as it changes across time and can vary in themes and ideals depending on social and economic contexts as well. For example, until the late 1800s it was rare for women in the United States to wear pants instead of dresses. By the 1970s it had become common for women to wear pants even in dressy situations. So the schemata of an individual from New York in 1850 compared to a New Yorker in 2010 in regard to what is normal in women's clothing would be very different, due to the formative social context.

Our formal and informal education systems also have a great deal of influence on the social context. Type of school attended—private vs. public, religious vs. secular, large vs. small, and so on—can also impact the context in which the schemata form, as can the ways in which topics such as history and social studies are presented. For example, textbooks are viewed in much the same way as scientists (as discussed in chapter 1), as infallible and truthful about our societal history, how we as a culture came to be. Yet in 2010 the Texas State Board of Education (one of the largest textbook purchasers in the country) voted to approve a social studies curriculum that requires altering the contents of history, sociology, and economics texts to present new views of history.

Among these changes was a reduction in the importance of the principle of the separation of church and state in the founding of the United States, a stipulation to use the term "free enterprise system" instead of "capitalism" in describing our current economic system, and a requirement to cut Thomas Jefferson from the list of figures who inspired eighteenth- and nineteenth-century revolutions, replacing him with St. Thomas Aquinas, John Calvin, and William Blackstone.[13] A student in a Dallas, Texas, junior high school in 1976 (the US Bicentennial year) and one in 2011 are going to get slightly different variants of our shared past, which will shape their perceptions of how of society came to be. Obviously, informal education at the familial and community levels is also important in the development of schemata. The views of parents and siblings, priests, rabbis, and imams, and the makeup of the community in which one grows up will also act to select just which parts of history and social values are presented to developing individuals.

Finally, it is critical to note that although I am making an argument for the impact of social context on overall schemata development, I am not suggesting that there is no suite of biological factors that help construct our behavior. In the section on physical development and bio-enculturation, I stressed the interplay between social and physical environments and biological systems. These biological systems are rooted in our genetics and developmental architecture and subject to a wide variety of constraints and patterns that affect how we act and think (see chapter 3). I am trying to explicitly focus on the entanglement between biology and culture here, to highlight the ways that cultural and environmental context shape our physical and cognitive development. There are human-wide patterns in biology and behavior that are important (many of these are discussed in chapters 4–6). I also do not want you to get the impression that schemata are uniform across a society; while aspects of them are universal, there is enormous variation in specific details. If you think about all the factors that we have outlined that go into affecting the developing schemata, you can see that each of us shares a good deal in common but has a relatively unique individual perception of the world as well.[14]

My point here is that the common assumption of humans having a basic nature (usually seen as biology) that is modified, or masked, via experience and environment (usually seen as culture) is too simplistic an explanation for why we are the way we are. In chapter 1 we saw that this concept of a basic human core with an overlaying of culture (and the role of free will) is pretty much the same whether you explain it as St. Thomas Aquinas did (via theology) or as Richard Dawkins does (via a particular atheistic kind of evolutionary perspective). Adhering to this bifurcated view keeps us susceptible to the major myths about human nature. What I hope to convince you of in the rest of this chapter, and the next, is that to better understand what it means to be human both culture and biology (evolution) matter equally, but not simply as two things added together to get a whole person. The rest of this chapter discusses what culture is and why it matters. Chapter 3 discusses what evolution is and why it matters. From these two discussions, we can assemble a tool kit to use to bust the three major myths about race, aggression, and sex.

CULTURE MATTERS . . . AND HOW

Because the word "culture" is going to be used throughout the book, we need to clarify how it is used and why it is important. In both popular and academic contexts "culture" can mean a variety of things. This leads to the question, What is culture?

People often use the term "cultured" to imply a certain kind of social status: popular arts (like action movies) are called low culture and other types of arts (like the opera) high culture. This is a hierarchical use of the term and refers to a specific society's categorizations (or valorizations) of different types of social and artistic activities. This is not the way the term "culture" is used in this book. In the nineteenth century societies were ranked according to specific expectations of what civilization and progress meant, but by the early twentieth century, anthropologists began to discuss the culture of societies in a different way. The most prominent nineteenth-century anthropological definition of culture was by E.B. Tyler: "Culture . . . is that complex whole which includes knowledge, belief, art, morals, law, customs, and any other capabilities and habits acquired . . . as a member of society."[15] Franz Boas, one of the most prominent early twentieth-century anthropologists, noted that people and their cultures interacted and shaped one another: "Culture embraces all the manifestations of social habits of a community, the reactions of the individual affected by the habits of the group in which he lives, and the products of human activities as determined by these habits."[16] In a key 1952 publication the anthropologists Alfred Kroeber and Clyde Kluckhohn reviewed 164 different definitions, synthesizing the range of concepts and offering what they hoped would be the seminal anthropological definition of culture:

> Culture consists of patterns, explicit and implicit, of and for behavior acquired and transmitted by symbols, constituting the distinctive achievement of human groups, including their embodiments in artifacts; the essential core of culture consists of traditional (i.e., historically derived and selected) ideas and especially their attached values; culture systems may, on the one hand, be considered as products of action, on the other hand, as conditioning elements of further action.[17]

Since then there have been a number of debates concerning culture in anthropology. The bottom line is that culture is what people do, think, make, and share. It is shared values and ideals, the symbols and languages, and daily patterns that make up our lives. Culture is both a product of human actions and a reality that influences those actions. We can think of culture as the shared, dynamic social context in which personal schemata develop.[18] But culture is not static. It changes over time and is not necessarily uniform in the ways individuals within a given society use and produce it. So for our purposes, culture is a core concept for understanding the development of how people perceive the world around them and also how those perceptions, ideas, and actions both perpetuate old and create new perceptions and behavior.

Culture and Meaning

The important aspect of culture is that it involves the realm of meaning generally. The fundamental characteristic of meaning is that it is established relationally via mutually reinforcing structures of significance; it derives from and is reproduced in social interactions with other people but also with and in particular social contexts.
—John Hartigan Jr. (anthropologist)[19]

Believing, with Max Weber, that man is an animal suspended in webs of significance he himself has spun, I take culture to be those webs, and the analysis of it to be therefore not an experimental science in search of law but an interpretative one in search of meaning.
—Clifford Geertz (anthropologist)[20]

Hartigan suggests that we see culture as a sphere in which meaning is created and provided to the actions and perceptions in our daily lives. Geertz sees the analysis of culture as one of the interpretations of the webs of significance we live in, create, destroy, and modify. It is the meaning of the cultural context (or web), for the people within it, that makes investigating it a powerful tool for understanding why people believe and act the way they do. When something happens, an action, an observation, and experience, our cultural context helps give it meaning, and our participation in that culture enables us to interact with that meaning, making the process dynamic and malleable. So, if culture has meaning then the symbols, ideals, and traditions we participate in come ready-made with relevance and connection to our personal schemata; they make sense to us.

Richard Shweder argues that a culture concept can be a beneficial analytic or abstract model to help us understand what people do; it is the community of perceptions transmitted within and across generations:

> By "culture" I mean community-specific ideas about what is true, good, beautiful, and efficient. To be "cultural," those ideas about truth, goodness, beauty, and efficiency must be socially inherited and customary, and they must actually be constitutive of different ways of life. (Richard A. Shweder, anthropologist)[21]

But if culture has meaning and helps us establish "sense" from action and experience, then its components are central players in the creation and maintenance (or eventual destruction) of the myths of human nature we are focusing on in this book; it enables us to perceive what is "good and right" specific to our historical and social context. Specifically, we are interested in the elements of culture that act to give meaning, the cultural constructs, the pieces of the web of significance that relate directly to misconceptions about race, aggression, and sex.

Cultural constructs are real

Generally, when we think about the elements of culture, we point to overt examples such as symbols (like flags or traditional clothing), ritual behaviors (such as those associated with religion, daily grooming, or greetings), linguistic patterns (forms of colloquial speech or slang), or specific social behavior (like personal space use and other aspects of body language). However, we do not always think about the cultural constructs—the ideas, ideologies, and systems of meaning—that pervade societies. A cultural construct is a belief, or social ideology, about the world that originates within a particular society and is (generally) shared by its members. That is to say, it is not necessarily tied to a specific event or a quantifiable and universally shared observation, but rather emerges from a conflux of historical and social elements and becomes widespread within the shared belief system of a culture. For example, the concept that the appropriate way to live as an adult is in a nuclear family (a married couple with children) in a separate residence from other such families, is a cultural construct common across much of American culture. This is not necessarily considered the preferred form of residence in all societies across the world (nor is it the most common form of residence or family unit); moreover, it is not based in some incontrovertible reality (biological or otherwise).[22] However, the ubiquitousness of this construct in the United States sets up a set of goals and ideals for both youth and adults and its attainment is often used as a gauge of social success.

All too often people make the assumption that because a cultural construct is "constructed" within a society that it is not a real thing. This is akin to people confusing psychosomatic ("it's all in your head") illness with fake illness. Whether your stomach hurts because of a bacterial infection or because of gastric disequilibrium due to stress from workplace tension, it still hurts. The mind can cause the body to react so as to create pain just as an infection can. The remedies might be different, but the pain can be the same. This is the case with cultural constructs; they are very real for those who hold them. In the example of the nuclear family and single-family residences, a failure to marry, have kids, and own a home can cause a wide range of problems socially and psychologically for individuals in the United States.

Cultural constructs are not necessarily stagnant. Gender roles are a good example. In the last century gender expectations have changed dramatically in the United States. In 1910 women played very few sports, rarely dressed in trousers, almost never managed or ran corporate businesses, and were considered mentally unfit to vote. By 2010 Serena and Venus Williams (tennis), Sheryl Swoopes (basketball), Michelle Wie (golf), and Danica Patrick (auto racing) were famous athletes, Hillary Clinton (former first lady) was secretary of state, Meg Whitman (former CEO of eBay) was running for governor of California, and Oprah Winfrey (media personality) ran one of the most successful companies in the United States. This is not to say that certain cultural constructs about women common in 1910 are not still at play today and that these women are not subject to them. For example, female athletes are often still presented as feminine sexual objects in addition to powerful athletes. However, it does show dynamism in the ways, and intensities, in which cultural constructs affect society and also that the constructs themselves can alter over time.

The beliefs and ideologies of a culture contain a number of cultural constructs and these in turn interact with the culture in a dynamic interface, setting up the social context in which humans develop physically and socially and their schemata form. However, not everyone in a given society shares the same experiences, contexts, and identical schemata. Certain cultural constructs are more pervasive than others. Beliefs about gender roles, aggression, and race are fairly ubiquitous and resilient cultural constructs are shared across a society. However, even within these general patterns there is variation in how they play out in different social contexts. In a society as large and diverse as the United States, there is a wide range of social contexts and patterns that affect the development and maintenance of schemata. Factors such as the region where one lives, the language or dialect one speaks, and one's ethnicity, religion, and socioeconomic class all provide somewhat distinct contexts and experiences that help shape our common sense notions of human nature and what we consider normal for humans to do.

Ethnic affiliation can impact people's political, economic, sexual, social, and religious views.[23] This reality is the core component in many reviews on the subject of ethnicity in the United States.[24] This is due to slightly different patterns of experience that characterize different ethnic groups (histories, political representation, residence patterns and geography, social and economic status, access to health care and education, etc.). These patterns are dynamic and have changed over the few centuries of this country's existence as the ethnicities vary, immigration and assimilation ebbs and flows, and cultural constructs about what is normal in the United States change. Beyond ethnicity, general socioeconomic status can also impact social context via access to goods and lifestyles as well as dietary and social and physical mobility constraints. Religions vary in their belief systems and thus different religions prioritize different aspects of shared cultural constructs; in addition, many have their own set of constructs that affect the development of schemata in their adherents. Even region of residence in the United States can impact social context. Different regions have different local histories. For example, the Civil War and history of slavery created somewhat distinct histories and social traditions in the northeastern and southeastern parts of the United States, and the long presence of Spanish, and then Mexican, culture and language before incorporation into the United States in the southwestern states and California shape what the cultural landscape looks and sounds like.

THE BASELINE: WHAT DO WE NEED TO KNOW ABOUT CULTURE FOR BUSTING MYTHS ABOUT HUMAN NATURE?

Culture is what people do, think, make, and share. It is the shared values and ideals, the symbols and languages, and daily patterns that make up our lives; it is the dynamic social context in which our schemata form. Culture is both a product of human actions and something that influences that action. So culture matters.

Here are the four core concepts about how and why culture matters, to use in our tool kit for busting myths about human nature:

1. Culture helps give meaning to our experiences of the world.
2. Cultural constructs are real for those that share them.
3. Individual schemata (our worldviews) vary depending of a range of elements in their social context.
4. Some cultural constructs are more pervasive or resilient than others, and thus more important to understand because they affect how we live and act and treat others.

The first concept is important because it helps us understand how we come with ready-made interpretations of what we experience and makes it easier to understand why many members of any given society share similar interpretations of events. But because culture is dynamic and we are all individually slightly different, there is some variation in those interpretations and thus the possibility of cultural change.

The second concept shows us that for members of a society, the beliefs and ideologies are not held as abstract concepts, but as things that are seen as real (or at least what is normal). This means that social and historical contexts give rise to particular ideas about reality that are firmly shared in a culture and interpreted as being true about the world, but are not necessarily biologically or historically accurate.

The third concept helps us understand why people in the same society might not see issues (and similar experiences) in the same way. In a broad sense this can help us understand why men and women might not perceive the same experience identically (gender roles are cultural constructs and very powerful in influencing the formation of schemata). This can also help us understand why different "races" or ethnicities (like "black" and "white" Americans) might not perceive a shared experience about discrimination, political history, or social equity in the same light.

The fourth concept is important because it helps us see that some beliefs and perceptions—especially those about race, sex, and aggression (the subject of this book)—are more ingrained within a society and resistant to change than others. The formation and propagation of these cultural constructs and perceptions are very important to understand because they have a major effect on how we live, act, and treat others.

Evolution Is Important—but May Not Be What We Think

Many have assumed that humans ceased to evolve in the distant past, perhaps when people first learned to protect themselves against cold, famine and other harsh agents of natural selection. But in the last few years, biologists peering into the human genome sequences now available from around the world have found increasing evidence of natural selection at work in the last few thousand years, leading many to assume that human evolution is still in progress.

—Nicolas Wade (science writer)[1]

This quote by a respected science writer illustrates how even the educated public may have a very poor understanding of what evolution is and what it is not. To sell the story, the false representation of a "debate" as to whether evolution happens in humans has to be a central theme. Two common misconceptions about evolution are revealed here: that it has an end point and that humans are somehow less affected by evolution than other organisms. The first sentence reports the common assumption that culture shields us from evolution and the second, that humans have made it, that we are already evolved.

Wade's *New York Times* article discusses current studies on specific parts of our genome (humanity's shared DNA system) that show evolutionary change in action. However, the popular concept that Wade refers to, the belief that we are insulated from evolution, oversimplifies the relationship between humans and their environments: assumptions that we are at the end of our evolutionary process are absolutely false. These ideas stem from a set of serious misunderstandings about what evolution is and how it works, and contribute to perpetuating the major myths we are going to bust in this book. Evolution is important, so we need to clarify a few main points before we can move on to tackling myths about human nature.

WHAT EVOLUTION IS NOT

When asked to explain evolution, most people would mention survival of the fittest, extinction of the dinosaurs, or mumble something about humans coming from monkeys. This reflects the reality that our education system and the consumer media do a very poor job of conveying the enormous amount of clear and reliable information we have about evolution and biology. Four common misconceptions about evolution will be discussed before we go on to describe what evolution is, how it relates to genetics and biology, and how we can use that information as part of our basic tool kit in busting myths about human nature.

The first misconception centers on the phrase "survival of the fittest." This is most often associated with Charles Darwin's theory of natural selection (one of the processes by which evolution occurs) and assumptions about evolutionary processes as the result of direct competition between organisms for a limited good (such as fighting over food or females with the bigger, faster, stronger individuals winning out).[2] This is the brute force of nature that so many philosophers and social thinkers have urged us to rise above, but it is not how evolution necessarily works.[3] Natural selection results in survival and reproduction of the sufficient or the good enough, not necessarily the best or most ferocious. The idea that nature is brutish and life is short owes more to philosophers such Thomas Hobbes and Adam Smith than it does to Darwin.[4] In fact, this concept is an important and powerful cultural construct that permeates our society.

A second major misconception about evolution is that it is oriented toward progress and results in organisms fitting perfectly with their environment. The corollary to this is that if something works well we perceive it as having "evolved" for this particular purpose. This is incorrect. Life on earth (and evolution) is messy and often haphazard. In some cases we do see amazing, almost perfect, relationships between organisms or an organism and its environment; for example, certain wasps and figs have coevolved and need each other to survive. However, in the vast majority of cases organisms work with each other and their environment in successful but much less than perfect ways.

This misconception of perfect coevolution is critical because it leads to the assumption that if something is a certain way, and it is the product of evolution, then it came to be that way for a purpose.

A third erroneous idea about evolution is that it all happens by chance. Because evolution is not goal directed, not manipulated by an external creative force, many people veer to the exact opposite perspective and assume that all evolutionary change is by chance. This is also incorrect. Evolutionary change is constrained by the materials at hand. That is, the current form of an organism and its evolutionary history (the history of change its ancestors underwent) affect the ways in which it can change in the future. Humans are not about to evolve wings for flight and tortoises' legs will not evolve into wheels, no matter what chance mutations arise. The principle that evolutionary change is constrained by the structure, development, and history of organisms is important and will help us understand why some of the myths about human nature are wrong.

Nearly three hundred years ago, before our current understanding of evolution developed, the author and philosopher Voltaire astutely noted a common trend in many Western philosophies, that the way that things are is the way that they ought to be:

> Observe that noses were made to wear spectacles; and so we have spectacles. Legs were visibly instituted to be breeched, and we have breeches. Stones were formed to be quarried and to build castles; and My Lord has a very noble castle; the greatest Baron in the province should have the best house; and as pigs were made to be eaten, we eat pork all year round; consequently, those who have asserted all is well talk nonsense; they ought to have said that all is for the best. (Pangloss, in Voltaire's *Candide*, 1759)

Pangloss's witticism, which mixes up "is" with "ought," illustrates the fourth misconception, that if something has evolved a certain way then that is the way it should be. This is the idea that the way things are has been selected for a particular purpose by evolutionary processes, or arisen by a sequence of chance events in nature, and thus must be the correct way (the "natural" way) for that thing to be. So, the beak of the hummingbird perfectly allows it to get nectar from flowers, the big brains of humans allow us to dominate the planet, and the eagle's feathers enable it to soar. However, none of these examples are decisive. There are many different kinds of nectar-feeding birds with an enormous array of beak shapes and sizes (although most are skinny and long); some other primates and many cetaceans (whales and dolphins) have brains roughly the same size (or larger) relative to their body size than humans do, and birds' feathers initially show up in their ancestors as thermoregulatory (heat-control) structures before there were wings (later feathers were co-opted and modified for use in flight). So all is not what it seems, nor is the status quo always for the best. A trait that currently serves a function for an organism may or may not have been altered by evolutionary processes to perform that function.

This fourth misconception is especially relevant to this book as it is often associated with cultural constructs and the mistaken habit of directly connecting biological traits with social behavior and social history. This is a main fault of Voltaire's Pangloss, and a common error in modern times. For example, there have been many attempts to argue that Americans of African descent are better suited for certain sports due to the biological effect and histories of slavery (seen as evolutionary selection) and that Jewish people, especially Ashkenazi Jews, underwent a series of events that provided them with higher intelligence (for evolutionary reasons) than other populations on average.[5] There are also numerous assumptions about men and leadership roles, as noted in chapters 2 and 6.

Finally, it is absolutely critical that we understand that evolution is ongoing . . . always. It is not about bigger, badder, and more beautiful winning the day; it does not stop at the perfect solution, nor is it goal directed. Most importantly, current aspects of our bodies and behavior that are affected by evolution are not more "natural" or correct than other parts of our lives and thoughts. So, having dispensed with what evolution is not, we need to clarify just what exactly evolution is.

Evolution is actually two things: a fact and a theory. When the majority of people say "evolution" they are usually referring to some facet of the theory, glossing over the most important part: the fact. The fact of evolution is that all organic life on this planet changes over time. Really, that's it: change over time. This is easily observable in the fossil record and in the laboratory. There is no debate, it is a scientific fact: evolution happens.[6] Now, there is a more specific definition of evolution based on genetics, but this is basically a more detailed way of saying things at the genetic level change over time. The theory of evolution is an explanation for how things change. A theory (as noted in chapter 1) is a set of hypotheses that have been tested and retested and supported by multiple researchers over time. The theory of evolution, then, is a set of well-supported ideas about how evolution works (like the theory of gravity). Currently we have a very robust theory of evolution that involves four main processes: natural selection, gene flow, genetic drift, and mutation. In the last few decades we have added a process called niche construction and a few other perspectives that involve new twists on the traditional four. We'll briefly explain each of these processes below, after we first set a baseline for a shared understanding of genetics.

Genetics: The Basics

As the shared unit of heredity across organisms DNA is core to understanding the processes of evolution.[7] To say that DNA, or deoxyribonucleic acid, is the main unit of heredity means that it is the way in which core information about organisms is passed from one generation to the next.[8] In sexually reproducing organisms, like humans, one set of DNA is provided by the mother and another by the father.[9] This DNA contains basic information that, when combined with the appropriate organic structures (in the egg) and context (the mother's uterus), will facilitate the growth of a single cell (the combined sperm and egg) into a multibillion-cell person. Note that I say "facilitate," not "determine." The DNA is not the blueprint of life, rather it contains many of the basic codes and signals for the development of an organism. At its core, DNA contains the basic information needed to assemble proteins, which are the building blocks of our bodies and systems, and it also acts to regulate how and where different proteins are made and used.

Parents pass down the information for the construction of proteins (as DNA) and the mechanisms for getting the development started (in the female egg) to their offspring. The specific stretches of DNA that contain these messages for proteins, or that regulate the building and use of proteins, are found in the same places in the DNA of all humans: they are what we call *genes*. A gene can be defined as a section of DNA that contains the sequence for a protein or the information for the regulation of some protein or proteins. All humans have the same genes, but for most genes there are numerous DNA sequences with slightly different chemical composition that can reside at the same segment of DNA. That is, each human has two copies of each gene (one from the mother and one from the father) but in the larger population there may be more than two forms of the gene.[10] These forms of the same gene are called *alleles* and are the basic genetic variation that provides the fuel for evolutionary change. The definition for genetic evolution is a change in allele frequencies over time. That is, the relative representation of specific forms of a gene can change across generations, which can lead to changes in proteins and eventually in physical traits.[11]

Although we can say that genes (and their alleles) contain the messages for proteins and their regulation, the actual relationship between genes and physical (or behavioral) traits is very complicated. There are complex chemical interactions inside our cells, interactions between cells, and developmental processes above the level of genetics throughout organisms across their life spans. This makes most one-gene-to-one-trait analogies unrealistic. For example, although your hands are composed of numerous proteins that emerge from messages in your DNA, hands themselves are not the direct product of a "hand gene." Hands are the product of a developmental program in which DNA plays an important, but not exclusive, role.

Think of genes as having four general types of relationships with traits (figure 1). First, a gene may simply help produce a protein (or several proteins if there is more than one allele for that gene). This is a simple "one gene–one protein" model. At a slightly more complicated level, we also get scenarios wherein a group of genes may work together to produce one effect, like a complex protein or a specific trait composed of multiple proteins (called a polygenic effect). Even more complicated, we also have situations where one gene can have many effects on a number of different traits and/or systems. That is, its protein product(s) affect multiple systems and targets (called a pleiotropic effect). Finally, it is very common for genes to have both polygenic and pleiotropic effects at the same time. Also, remember that each gene can have more than one allele, so in all of the above cases the same gene can be producing slightly different proteins in different individuals.

Multiple factors influence the development of an organism. We can think of two facets of an organism, the genetic component, or *genotype,* and all the traits of the organism, the *phenotype.* Major physical traits (parts of the phenotype) like body size and shape, head shape, face form, and so on emerge from the interaction of many genes and specific developmental and environmental influences. These include chemical and physical patterns, internal and external influences, and physical constraints on shape and size, in addition to the messages and regulatory processes carried in the genes. To make things even more complex, starting with conception (the successful joining of sperm and egg), epigenetic (outside of the DNA) developmental processes also affect development. Changes in temperature, fluctuating chemical environments, and mistakes in chemical cues in

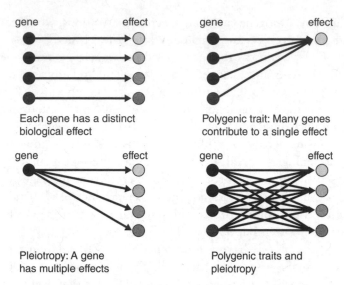

Each gene has a distinct biological effect

Polygenic trait: Many genes contribute to a single effect

Pleiotropy: A gene has multiple effects

Polygenic traits and pleiotropy

FIGURE 1. The types of relationships genes may have with traits. The most common relationship is also the most complex: a combination of polygenic and pleiotropic patterns. Adapted from A. Fuentes (2011), *Biological Anthropology: Concepts and Connections,* 2nd ed. (Burr Ridge, IL: McGraw Hill Higher Education).

addition to variations in DNA produce slightly different outcomes for each organism.

What about genes for certain behaviors, tendencies, or abilities? Currently, there is little evidence to support any one-to-one relationship between a single set of alleles and any behavior. However, DNA does influence our physical structures (brain, eyes, mouth, hands, and so on), and since behavior is exhibited via these structures, all behavior has some genetic component. Take an example I use for a basic textbook explanation of this.[12] You are reading this page using your eyes (optical tissue, muscles, nerves) and maybe your hands (muscle, bones, tendons) to scan the letters and words on the page. You are also using your brain (a set of neurons, vascular tissues, and various hormones that connect all the organs in your body and mediates among them) to process the meaning. All of these elements have a genetic component. However, you are reading the words, a behavior that must be taught to you, and you are reading them in English, something else that must be taught to you. Do reading and using the English language have a genetic component? Yes, the neurons, eyes, muscles, and other parts of the body used in reading are composed of proteins initially coded for by DNA. Are there genes (or alleles) for reading in English? No, the specific language that someone reads is an experiential factor, as languages are parts of cultural systems. Can aspects of our genetic complement impact our ability to acquire specific reading skills? Possibly. Structural differences in the eyes, motor connectivity, and even hormone pathways in the brain might impact the pace and pattern of reading acquisition. These differences could be the result of allelic variation or other genetic patterns.

So trying to connect genes and behavior is not at all simple. We know there are genetic components to all aspects of life, but how or whether specific alleles affect complex behavior patterns is far from obvious. If you do hear an assertion about a relationship between genes and behavior, you need to think critically and ask a few basic questions. What is the gene or genes? How many alleles are there? What protein or proteins are coded for? How do these proteins affect the organism so that a specific behavior is performed? As our technological abilities advance, we may be able to dissect the mechanism of DNA function as it relates to behavior, because we know it must. But for the present we need to acknowledge the limits of our current understanding and carefully assess any such claims.[13] There is a very complex set of relationships between the phenotype, or end product, on the one hand, and DNA, development, and environment on the other. This relationship is not linear, nor can it be easily described as a simple equation. We should not use simple models or labels such as "blueprints," "building blocks," or "code of life" to describe DNA and genes. Rather, the DNA is an integral component of life itself, and understanding the units of heredity and the function of genetic material is critical to understanding evolution and the functioning of organisms. But an understanding of genetics is by no means the complete picture. By combining our understanding of genotype and phenotype, we can better understand how evolution changes populations over time.

No two individuals are identical. Even when the exact same segments of DNA are involved (such as in clones of identical twins), there are still many complexities in the ways that genes are expressed. Genes may have multiple types of effects, developmental patterns and changing or shifting environments can alter the outcomes of development, and mutation can occur. The sheer number of variables that go into the expression of genes and the development of organisms ensures that no two individuals, even genetic copies such as clones or identical twins, will be the exact same people. However, in spite of all this variation people do still tend to look like their parents; also, members of a population often resemble one another. This is due to the amazing ability of the genetic system to regulate and correct errors and keep the overall pattern of development reasonably consistent, and the fact that evolutionary processes often tend to keep individuals within populations relatively similar to one another.

A basic understanding of genetics tells us that we are simultaneously remarkably similar (all having the same genes) and yet individually diverse (due to variation in the alleles people have) at the same time. We also know that our genes play a role in our development and functioning, not as directors, but rather as part of a complex system. Understanding evolution entails understanding variation at the level of individual genetics and population-level genetics. We need to understand how variation arises and how populations change or stay the same over time, and how specific variants (genetic or general traits) become more or less common over time. These basic concepts convey a core understanding of how evolution works.

Evolution: The Basics

Our current understanding of how evolution works contains four main components: mutation, gene flow, genetic drift, and natural selection. The first component, mutation, acts to create new variation and the other three move that variation around and shape populations.

Mutation is the only way to get new genetic material. Basically, mutation refers to a chemical change in the DNA. There are a number of ways this can happen, and usually it is neutral (meaning these changes are invisible outside the DNA). This is because DNA has a series of self-repair mechanisms and the messages in the genes are designed to be redundant, so that small chemical changes often do not impact the actual protein or regulatory product of a gene. However, sometimes they do, often disrupting a gene's function (altering the protein production or regulation) and causing problems for the organism. However, every so often a mutation results in a new sequence (a new allele) in a gene that actually produces a better-functioning protein or regulatory process. The bottom line is that mutation occurs for a number of reasons (many of them random with respect to location and pattern), usually with neutral or negative results, and sometimes with beneficial results.

It is important to note that mutation generally results in small changes in the chemical structure of DNA, not giant changes in an organism. Major changes in the shape or function of the whole body or major organs are due to errors in development, not genetic mutations (usually).[14] These larger developmentally induced changes might be heritable, but usually are not. One can think of changes to the fetus due to factors affecting the mother's body or possibly an error in the development of certain tissues. A dramatic example would be when a genetically male (XY) individual has a developmental error in the areas of hormone receptors and the testosterone flood at the sixth week of fetal gestation fails to trigger the development of testicles and a penis. This individual will develop with more female appearing genitalia (this is not heritable, but does tell us a lot about development of the body).

One of the most overlooked processes of evolution is gene flow. Very simply, gene flow is the movement (or lack of movement) of alleles within and between populations. Remember that within a large population there may be multiple alleles for each gene (but each individual can only have a maximum of two different alleles per gene) so each population has its own pattern of representation of specific alleles. For example, take a species with gene A and three alleles, A1, A2, and A3, each producing a protein that affects hair color in a slightly different way (A1 proteins make hair lighter and A3 make it darker, with A2 intermediate). Population 1 has the following allelic representation for gene A: 80 percent A1, 15 percent A2, and 5 percent A3; this population's hair is relatively light. Population 2 has the following allelic variation: 17 percent A1, 20 percent A2, 63 percent A3; individuals in this population have relatively darker hair. Note that in each of the populations there is variation, with all hair types present, just at differing frequencies for the underlying alleles. If there is no gene flow between the populations then these frequencies might stay the same (more or less). However, if there starts to be regular gene flow (migration) between the populations then we might see the patterns of allele frequencies start to balance out with more A3's coming from population 2 to population 1 and more A1's moving from population 1 to population 2 as individuals from the two populations mate. The result would be that relative frequencies of hair color in the two populations would become more similar. So gene flow acts to make populations' allele frequencies (and any phenotypic correlates) more similar, and lack of gene flow restricts the potential for populations to share the same allele frequencies.[15]

Genetic drift is the one truly random process of evolution. Basically, the premise is that events that alter allele frequencies in populations happen randomly sometimes. For example, in small populations natural disasters like earthquakes or floods can take out a chunk of a population with some individuals surviving and others not, simply due to chance. This could radically alter the genetic makeup of a population. Using the hair color example, say a small population of the organism with the hair color gene A live on an island. Their allelic frequencies are 80 percent A1, 10 percent A2, and 10 percent A3. It turns out that most of the population, including most of those with the A1 allele, lives on the north coast of the island (by chance) and a major tsunami hits, wiping it out. The remaining population comes from the small groups on the south end of the island, where the A1 allele is not as common. So after the tsunami the population has a new relative set of alleles frequencies with A1–3 all at about 33 percent. Evolution occurred, but not for any reason having to do with the organism or the alleles in question. Drift is most pronounced in small populations because in very large populations, for every random change in one direction there is likely to be a random change in the next.

The process called natural selection is what most people think of when they hear the word "evolution." Natural selection is the process for "descent with modification" first proposed by Charles Darwin and Alfred Russel Wallace in the mid 1800s.[16] Since then we discovered DNA, made great advances in understanding genetics and developmental systems, and have a much better idea than Darwin could ever have imagined with regard to the process he envisioned. At its most basic the premise of natural selection is as follows:

1. There is biological variation in living forms.
2. Some of this variation can be passed from one generation to the next.
3. Within any given environment some variants help the organism leave more offspring than others (on average).
4. Those variants that help organisms do better, if they are heritable, will over time become more common (because more of the next generation's offspring come from parents having those traits).
5. Those variants that become more common in a population are seen as adaptations to the particular environmental contexts.

This is fairly easy to understand given what we have already covered about development and genetics. We know that each individual in a population is slightly unique and that for most genes or traits there are at least a few variants in any population. If in certain contexts one form of a gene (actually its protein product's effect or its regulatory effect) makes it slightly more likely that those individuals having that allele will survive and reproduce (pass the allele on to the next generation), then over many, many generations the population will have more of the allele that provides the slight reproductive advantage and the allele frequencies will have changed over time (evolution has happened).

However, there is a slight problem with this simple version: DNA does not generally interact directly with the environment. It is the whole organism (the phenotype, or collection of all the traits) that interacts with the environment. Phenotypes of organisms pass through a sort of filter in each generation, with differential success in leaving copies of their genotype (the genetics) in the next. This suggests that primarily those parts of the phenotype (traits) that are somehow linked to genetics are subject to natural selection (though this is not the whole picture). So if phenotypes are packages (collections of traits), and it is the entire package that interacts with the environment, it is the collective effect of a given filter on the whole phenotype that sets up differential success among individuals in passing along DNA in that population.

At a basic level natural selection emerges from phenotype-environment interactions in which some phenotypes do better, on average, than others in a given environment, and thus the genetic basis (genotype) for these phenotypes is increasingly represented in subsequent generations of the population. It is critical to see that "success" in natural selection simply means leaving more surviving offspring, on average, per generation. Natural selection is not a life-or-death battle between individuals in a population in every generation; it is a long, drawn-out series of interactions in which a slight reproductive advantage changes the genetic makeup (genotypes), and thus the physical makeup (phenotypes), in a population over the long run.[17] Specific traits (parts of the phenotype) that do best in a given environment are called *adaptations*.

For example, when we say that the big colorful tail of a peacock is an adaptation, we are stating that this is a trait that has become more prominent over time because it confers on peacocks with it better chances of leaving more copies of their DNA in subsequent generations. This is a good example of the complexities, even at this simple level, of natural selection. While having a big colorful tail makes a peacock more susceptible to predation (being eaten because he is slow and colorful) it also seems to help him attract peahens and thus mate effectively. Such a trait can be an adaptation if, on average, the owners of flashy tails leave more offspring than those with less impressive tails. Note that this is not the nineteenth-century English poet Alfred, Lord Tennyson's description, "nature red in tooth and claw," with the strongest or fastest or even the prettiest surviving. Rather, a range of tail traits "survive" in the population of peacocks, but over time the flashier ones tend to become more and more prominent in the population of the species. This leads us to another facet of natural selection: it tends to decrease variation over time to the narrow range of those traits that do well. For example, all humans walk on two legs in pretty much the same fashion. While there is a lot of variation in minor details of how people walk (and some interesting differences between males and females), natural selection has winnowed the variation in how we walk because it is quite important to our basic functioning that we are effective at walking on two legs.

This brings us to the core component of natural selection: it is evolution related to function. Gene flow and genetic drift do not necessarily have anything to do with function; they cause changes over time without any regard for what the traits or alleles that they affect do. Not so with natural selection. The mode of change for natural selection is environmental filtration across generations for better-functioning traits or genetic complexes and thus outcomes that are tied to what traits do. So traits that arise from natural selection are called adaptations because they were shaped over time by the "fit" between what they do and the environments in which they exist. This is actually what is meant by "survival of the fittest," that those traits that become most common due to natural selection are those that fit best in a given environment. However, what makes for a good fit in one environment might not work in a different environment. We are not talking about survival in the sense of living, but rather in the sense of representation across generations in particular environments.

Mutation introduces variation, gene flow and genetic drift mix that variation around, and natural selection shapes the variation in relation to specific constraints and pressures in the environment. Sometimes these processes can be at odds with one another, but they also work together. Remember, there is no goal for evolution. Each generation simply responds to the current pressures with whatever variation is available at that time and passes on some subset of that variation to the next generation. Over time (many, many generations) the range of variation and the types of pressures shaping the way populations look vary depending on the environments and the interactions of individuals with the environment. This is a very basic description of the core processes of evolution. However, to better understand the relationship between humans and evolution in relation to our project of busting myths we need to add some newer insights about evolution.

Probably one of the most important major innovations in evolutionary theory in decades is the concept of niche construction. Formally proposed by the biological anthropologist John Odling-Smee and the biologists Kevin Laland and Marcus Feldman, niche construction is the concept that organisms and their environments interact with and shape one another across evolutionary time.[18] Under the traditional assumptions about natural selection, the environment acts as a filter and constraint on organisms, causing functional changes over time. When you add niche construction, organisms gain an active role in shaping the environment in response to the pressures it exerts. This leads to an organism-environment interaction that is dynamic rather than static. Take earthworms for example. When you introduce earthworms to a new plot of soil they are challenged by the soil and its chemical consistency. Over time, as the worms move through the soil they ingest it and excrete it (that is how they feed) and that process alters the chemical makeup of the soil and its consistency, changing the pressures that the soil environment exerts on subsequent generations of earthworms in the same location. Other examples include beaver dams, multigenerational nest building in birds, and even things like tool use in apes. Basically, niche construction results in an ecological inheritance (resulting from previous organism-environment interactions) that goes alongside the genetic inheritance. This results in a more complex set of organism-environment interactions than originally envisioned via basic natural selection. Of course, it should be obvious that above all else, humans engage in substantial niche construction affecting the ways we can interact with the environment:

> Niche construction theory may be particularly relevant to the dynamics of cultural traits as the theory can incorporate the effects of the cultural background as a form of constructed niche. (Kevin Laland, biologist; Jeremy Kendall, anthropologist; and Gillian Brown, psychologist)[19]

Laland and colleagues point out that inheriting ecological contexts via human material culture (tools, clothes, towns, etc.), and niche construction in general, can occur via cultural means. In this case humans can be seen as the ultimate niche constructors and cultural processes can provide a particularly robust method of niche construction. So niche construction plays a key role in human evolution via our ability to modify our surroundings through behavior. Just think about agriculture, building houses and towns, and extracting minerals from the ground to make iron and steel. It is important, then, to see that evolutionary processes (natural selection and gene flow especially) can be modified as they occur via niche construction.

This idea that human behavior impacts evolutionary pressures is also invoked by proponents of gene-culture coevolution.[20] Here it is argued that cultural innovation and activity act, in a sense, as a part of the environment and as a mechanism for niche construction that influences how natural selection and gene flow interact with human populations. Think about the advent of agriculture and village life and how those cultural aspects altered the pressures of natural selection and the patterns of gene flow (which in turn would impact the genetic and phenotypic traits in populations). Agriculture relieves the stress of getting wild foods but requires a different kind of time and energy, plus a social commitment to stay in the same place for long periods of time. Villages gather large groups of people together, which can help in growing crops and defending the group, but at the same time may radically increase the risks of transmitting disease.

Finally, another important addition to our understanding of evolutionary processes is the concept of inheritance systems beyond the DNA, proposed by the biologists Marion Lamb and Eva Jablonka.[21] They argue for recognition of "evolution in four dimensions" rather than a focus on just one—their main point being that practitioners of traditional evolutionary approaches focus on just one system of inheritance, the genetic. Jablonka and Lamb argue for a new perspective by including three other inheritance systems that may have causal roles in evolutionary change: epigenetic, behavioral, and symbolic inheritance.

Epigenetic inheritance is found in all organisms, behavioral inheritance in most, and symbolic inheritance occurs only in humans.[22] While not negating the importance of genetic inheritance in the evolutionary process, this concept of multidimensional inheritance, like niche construction and gene-culture coevolution, points out that in many organisms (especially humans) behavior and alteration of the environment can play major roles in the way biological change occurs over time. Of particular interest here is the concept that symbolic inheritance (often in the form of cultural constructs) can impact biological change via the behavior and action of individuals.

Our basic understanding of evolutionary theory (how evolution works) in the early twenty-first century may be summed up as follows:

1. Mutation introduces genetic variation, which may introduce phenotypic variation.

2. Developmental processes can introduce broader phenotypic variation, which may be heritable.

3. Gene flow and genetic drift mix genetic variation (and potentially its phenotypic correlates) without regard to the function of those genes or traits.

4. Natural selection shapes genotypic and phenotypic variation in response to specific constraints and pressures in the environment.

5. At any given time one or more of the processes above can be affecting a population.

6. Dynamic organism-environment interaction can result in niche construction, changing pressures of natural selection and resulting in ecological inheritance.

7. Cultural patterns and contexts can impact gene flow and the pressures of natural selection, which in turn can affect genetic evolution (gene-culture coevolution).

8. Multiple inheritance systems (genetic, epigenetic, behavioral, and symbolic) can all provide information and contexts that enable populations to change over time or avoid certain changes.

REVISITING THE MISCONCEPTIONS ABOUT GENETICS AND EVOLUTION AND BIOLOGY

Given what we've reviewed in this chapter, it behooves us to revisit a few of the core assumptions about genetics and evolution that help prop up the myths of human nature that we will be busting. The first and most often incorrectly invoked is the concept of survival of the fittest. This refers only to the outcome of natural selection, not the other processes of evolution. So right away we see that this is a limited notion of what evolutionary change might produce. Most people consider this concept as meaning that there is a struggle for survival between individuals, and those who are "better" (usually stronger, faster, smarter, etc.) will survive the competition and pass their genes into the next generation.

This is not the way natural selection works. In an evolutionary sense being "fit" means having a trait, or collection of traits, that, in a certain environment, can provide an individual with a better chance of passing their DNA into the next generation (on average) relative to other individuals lacking those same traits (or the same variants of those traits). Much of the selection results not from direct competition between individuals, but rather from the environmental filter that populations go through in each generation.

This is much less exciting and less competition-based than the common perceptions of natural selection. If there is substantial variation in a specific trait, and that trait is heritable, then we can see selection favoring certain patterns over time. However, if the trait is less variable, or there are a number of variants that do well, then we will not see much differential selection between individuals at all. For example, in humans today the trait of walking on two legs (called bipedalism) is ubiquitous. Despite there being a lot of variation in how people walk, most people walk well enough to not have it inhibit their chances of reproducing. There are many traits shaped by selection in the past that retain variation, but all the variants are sufficiently effective so selection is no longer a major pressure on the existing variation.

The second clarification brings us back to Volatire's *Candide* and the assertion by the fictional philosopher Pangloss that we are in the best of all possible worlds and things are as they are supposed to be; this assertion is powerful if we want to assume that what we see as "evolved" (produced as a result of evolution) is the best state for that thing to be in. This assumption is relatively noncontroversial when we think about legs and walking, but a lot more complicated when we think about gender roles and how men and women seek out and participate in sexual activity. If we invoke an evolutionary explanation for something (such as in issues related to race, aggression, and sex), we often equate that with being in sync with nature. It is one of our society's cultural constructs to assign a certain intrinsic value to things we deem natural (coming from outside of human control) as opposed to things we see as being altered by human society and customs. It is easy to argue that walking on two legs is natural for humans, but walking on our hands is also natural (our hands arose via similar biological developmental processes as our feet, but just not as practical as a way of regular walking). The point is that we have to exert extreme caution when arguing that something is "natural," or the way it should be. Assuming something is natural does not necessarily mean it is "fit" or "correct." Nor does it necessarily mean that social and historical contexts did not have a hand in shaping it. This specific issue will be very obvious in some of the core assumptions for each major myth we are going to bust.

A third clarification concerns what people think genes do, what genes really do, and the relations between genes, traits, and behaviors. Many people think genes are the main cause of who we are, what we look like, and how we behave. This is called genetic determinism, and it is largely wrong. However, this does not mean that our genes are not importantly involved in our bodies and behavior, just not primarily in a causal way. The bottom line is that genomics is complicated, but we are all able to understand the basics. Genes contain the information for the production of proteins and, in some cases, information for the regulation of genetic functioning. Most genes come in multiple forms that can produce slightly different proteins or regulatory actions. Most gene products act together with other aspects of the body during development and throughout the life of the organism. Changes in the relative frequency of different forms of genes in populations is one of the ways we measure evolution in action, and this can help us understand why there are differences between populations of humans. So, genes are important in understanding humanity, but they are not the blueprints, or cause, for our natural selves. We must also remember that DNA is not the only form of inheritance. There are development patterns that are inherited and an array of nongenetic factors (ecological, cultural, and historical) that are inherited and not tied to our DNA.

In regard to the relationship between DNA, development, and our phenotype, we have to remember that many traits (both in the structure of our bodies and in our behavior) are considered neutral relative to natural selection. If there is not a clear connection to changes in the rate or efficiency of passing along DNA from generation to generation, then natural selection does not even "see" the trait. This means that these traits are not considered functional from an evolutionary standpoint. So, much of the variation in traits we see in humans is not distributed due to selection, but rather due to processes of change that are not related to evolutionary function. A good deal of what we see as biological variation in humans (which we usually consider "natural") is neutral. Take for example one of the defining characteristics of modern humans, our chin. This protuberance on our lower jaw is one of the few clear characteristics that separates us from earlier forms of humans (like Neanderthals), but it is a byproduct of selection on other parts of the jaw and skull. The chin, as a trait, is neutral and not a focus of selection, but yet it remains a diagnostic trait in defining humanity. The point that biological traits may be neutral in regard to selection will prove crucial to busting certain myths about human nature.

Finally, understanding what evolution actually is, and how genetics relates to it, gives us a better ability to assess claims about what forms and functions have emerged from human biological histories and how those might relate to the basic patterns of being human. We can avoid falling into the "evolved for" trap and step back and ask effective questions about how humans vary and how that variation might be related to our biological and evolutionary histories.

THE BASELINE: WHAT DO WE NEED TO KNOW ABOUT EVOLUTION AND GENETICS?

Genetics and the structure of our bodies are major factors in making up who we are and how we do what we do. The processes of evolution influence genetics and the development and functioning of our bodies, and interact with cultural, ecological, and historical facets of humanity to comprise a major component for understanding how humans look and behave. Indeed, evolution matters.

Here are the core concepts about evolution to use in our tool kit for busting myths about human nature:

1. Evolution is change over time. Specifically, it is change in genotype and phenotype across generations due to a variety of processes.

2. Mutation generates new genetic variation. Gene flow and genetic drift move that variation around, and natural selection shapes the variation in response to environmental pressures.

3. Niche construction theory suggests that humans and their environments are mutually interactive participants in the evolutionary processes through ecological inheritance. Multiple inheritance theory illustrates that evolutionarily relevant inheritance can be at the genetic, epigenetic, behavioral, and symbolic levels.

4. Our DNA alone does not determine who we are and how we behave, but it is a primary component in the development and maintenance of our bodies and behaviors. Genes contain the basic instructions for the building blocks (proteins) of biological

systems. Genes and our phenotype are connected, though not usually in a one-to-one relationship, and are influenced by evolutionary processes.

The first concept is important because it clarifies what evolution is, a necessary baseline for asking any question about human nature. The second concept is important because it is the traditional understanding of how evolution works, and the third concept adds critical new perspectives, making our abilities to ask questions about humans more robust. Finally, the fourth concept reminds us that we should not be looking for primarily gene-based explanations for being human, but when constructing new explanations, or busting old ones, we must keep genetics and the biological facets of evolutionary processes as core components of our explanatory tool kit.

PART TWO

Busting Three Myths about Being Human

Human ≠ Nature + Nurture

As theoretical possibilities, one can envisage that man
might be genetically determined as aggressive or submissive,
warlike or peaceful, territorial or wanderer, selfish or
generous, mean or good. Are any of these possibilities
likely to be realized? Would the fixation of any of these
dispositions, so that they become uncontrollable urges or
drives, increase the adaptiveness of a species which relies on
culture for its survival? I believe that the answers to these
questions are in the negative.

—Theodosius Dobzhansky (evolutionary biologist)[1]

A man should look for what is, and not for what he thinks
should be.

—Albert Einstein (physicist)[2]

Being human is messy. We are simultaneously biological and cultural
beings with complex schemata and social lives that shape and populate
our perceptions and philosophies: we are naturenurtural.[3] As Theodosius
Dobzhansky notes, we have not evolved to have one particular way of
being human; there are a number of potential outcomes to the human
experience. However, the way we see the world and our cultural
inheritance limit the ways in which we can perceive and experience
those potential paths. Albert Einstein urges us to make an attempt to
see what is really out there as opposed to seeing only what we already
believe to be there. These quotes from Dobzhansky and Einstein set the
stage for the next three chapters, which bust three myths about who we
have evolved to be and illuminate what is actually there in the human
experience of race, aggression, and sex.

A baseline understanding of culture, genetics, and evolution makes clear that human beings are extremely complex and requires us to dump the nature/nurture concept in order to tackle the "Big Three" myths about human nature. We are not a blank slate at birth, to be filled in via cultural experiences; rather we are born as an organism, a collection of organs, tissues, and cells generated by the interactions of DNA and all of our developmental processes, which in turn have been shaped by our evolutionary histories. As a human organism we are born into a suite of inherited ecologies, cultural patterns, and social contexts that immediately become entangled with our biological structures, initiating a process of biocultural development: we are naturenurtural.

Some of our biological structures develop in particular ways largely due to the individual patterns of our DNA and our biological histories—things like hair color, nose shape, and the shades of our skin. However, these structural parts of our body exist within the cultural context we are a part of. So we may shave or color our hair, surgically change the shape of our nose, and alter the shades of our skins; we are active in the way we live in and perceive our bodies, and how we view the bodies of others. Our nose may be a collection of tissues arising from the interactions of genetics and development, but for every human a nose is just as much part of our cultural selves as it is part of the structure of our face.

Other aspects of being human, such as our tastes in food, who we are attracted to, the sports we play, and how well we play them emerge from the mutual, and interactive, development of our bodies and our experiences. Whether it is our adult height and weight, our ability to score well on standardized exams, the views we have on raising children, our resistance to disease-causing bacteria, or what we think of as natural behavior for a man or a woman, each of these is a product of different relationships that meld the biological and the cultural, the historical and the evolutionary, into a single process.

The big three myths about human nature are so prominent because they rely on our tendency to assume that culture plus biology equals us. Becoming human is not a simple addition problem. One way to envision this is via the concept of potential versus performance. Think of *performance* as the expression of any given trait (physical or behavioral) and *potential* as the underlying variation and constraints (genetic, physical, and cultural) that affect the range of possible performance. For a very basic physical example, take the length of your femur (the long bone in your upper leg). In the fetus the mass of tissue that is eventually going to be the femur is set up largely by genetic action and protein interactions. During fetal development the general shape of that bone is influenced by interactions among cells and between different masses of tissues and hormones (which can be also influenced by the mother's health and behavior). After birth the rate of bone growth is heavily influenced by the type of nutrition received, the activity patterns of the growing child, and disease exposure. So you could have two individuals with almost identical potential spectrums for femur length, like twins separated at birth, and different experiences during development could result in different actual femur lengths. For a behavioral example, consider the exhibition of physical aggression in humans. There is a huge range of potential types and patterns of physical aggression that any individual human can exhibit, but the actual performance of physical aggression is influenced (minimally) by body size, muscle density, gender, health, cultural patterns, life experience, and the availability of weapons or other tools.

Revisiting the theme from the Dobzhansky and Einstein quotes we can think of the myth-busting exercise as an attempt to find out what the actual details are of human potential and performance related to race, aggression, and sex, *and* why this reality does not necessarily line up with peoples' perceptions about them. Remember that in chapter 1 the philosopher Mary Midgely cautioned that "we are accustomed to think of myths as the opposite of science. But, in fact, they are a central part of it, the part that decides its significance in our lives. So we very much need to understand them. . . . They are imaginative patterns, networks of powerful symbols that suggest particular ways of interpreting the world." Our notions of what is natural, of what science tells us about humankind, and of what we see and accept as the reality of race, aggression, and sex are influenced by these myths and color how we see the world and act in it. Because myths are powerful and relevant to our everyday lives it is important to challenge these myths' assumptions about human nature.

But why are the big three myths so important to bust? The fallacy behind the race myth is that humans are divided into biological races (black, white, Asian, etc.) and that there are certain natural differences among these groups. If we believe this to be true, it shapes the way we act toward and perceive others, what we expect, and what we think we can achieve, as far as human equality, and whether or not we can build community in an increasingly diverse society. The fallacy behind the myth of aggression is that nature and nurture are different, and that our animal (or evolutionary) core is that of a primitive beast. If this were true, then our nurture (cultural constraints) would manage an inner nature (a primitive, aggressive drive especially in men) that emerges whenever the grasp of civilization weakens. If we believe this then we will accept a wide range of interpersonal violence as inevitable, and we will see war, rape, and murder as just part of the nature of human beings. But if these are all part of our potential, and not our nature, then a much broader range of responses and ways of living together become possible. Finally, the fallacy behind the myth of sex is that men and women are truly different in nature, and that this difference emerges in our behavior, desires, and internal wiring. The relationships between, and within, the sexes and genders are constrained by such a view, and the possible range of ways to be and become human and express our sexual and social selves is extremely limited. If differences in the sexes are present but less extreme and occur in different ways than we currently envision, the possibilities for human relations expand and, as in the two previous cases, our abilities to build communities and coexist in sustainable ways in our increasingly crowded, diverse, and complex world could become slightly improved.

Busting myths of human nature is not like busting a myth about the quality of air on an airplane or the effectiveness of vitamin C at preventing a cold. There is generally no one single test to refute an entire myth and there is usually not even a simple right or wrong answer when assessing the parts of the myths. There are multiple points and complex sets of data and theoretical concepts. Busting myths about human nature is neither flashy nor easy—it is very complicated, so bear with me. In chapters 4–6 I will try to walk us through the myth-busting sequence, starting with the myths and then slowly and thoroughly showing how the available data challenge, modify, or refute the assumptions underlying the core parts of each myth. These chapters are each organized around three topics. The first topic is an outline of the extent of the core myth and its premises, the second topic examines four specific subsets of the myth that we will bust, and the final topic tries to pull all the information together to present a more accurate view of the myth (race, aggression, or sex) based on the broadest current information available.

Here are the eight basic concepts—the tool kit—developed in chapters 2 and 3; keep them in your mind as we tackle myths of race, aggression, and sex in the next three chapters:

1. Culture helps give meaning to our experiences of the world.

2. Cultural constructs are real for those that share them.

3. Schemata (our worldview) vary depending of a range of elements in their social context—this explains why people in the same society might not see issues (reality) in the same way.

4. Some constructs are more pervasive than others, and thus more important to understand as they affect how we live and act and treat others.

5. Evolution is change over time. Specifically, it is change in genotype and phenotype across generations due to a variety of processes.

6. Mutation generates new genetic variation, gene flow and genetic drift move that variation around, and natural selection shapes the variation in response to environmental pressures.

7. Niche construction theory shows us that humans and their environments are mutually interactive participants in the evolutionary processes and helps us realize that ecological inheritance is important. Multiple inheritance theory illustrates that evolutionarily relevant inheritance can occur at the genetic, epigenetic, behavioral, and symbolic levels.

8. Our DNA alone does not determine who we are and how we behave, but it is a primary component in the development and maintenance of our bodies and behaviors. Genes contain the basic instructions for the building blocks (proteins) of biological systems. Genes and our phenotype are connected, but usually not in a one-to-one relationship; however, the relationship they do have is shaped and influenced by evolutionary processes.

4

The Myth of Race

The idea of "race" represents one of the most dangerous
myths of our time and one of the most tragic. Myths
are most effective and dangerous when they remain
unrecognized for what they are.

—Ashley Montagu (anthropologist)[1]

Ashley Montagu, one of the most prominent anthropologists of the
twentieth century, warned about the pernicious myth of race in 1942,
and his warning is still relevant today. In his 2010 book, Guy Harrison
challenges the biological reality of race:

> Few things are more real than races in the minds of most people. We are
> different. Anyone can see that. Look at a "black" person and look at an
> "asian" person. If a black Kenyan stands next to a white guy from Finland
> we all can see that they are not the same kinds of people. Obviously
> they belong to different groups and these groups are called races, right?
> (Guy Harrison, journalist)[2]

Guy Harrison is calling into question the most common popular
perception of human variation—that if we can see differences, if we can
tell people apart, then there must be real (meaning natural) differences
between groups of people.[3] The question of whether humans are divided
into biological races is answered with a resounding academic "no" by
the American Association of Physical Anthropology's (AAPA) statement
on the biological aspects of race:

> Humanity cannot be classified into discrete geographic categories with
> absolute boundaries. Partly as a result of gene flow, the hereditary charac-
> teristics of human populations are in a state of perpetual flux. Distinctive
> local populations are continually coming into and passing out of existence.
> Such populations do not correspond to breeds of domestic animals, which
> have been produced by artificial selection over many generations for spe-

cific human purposes. There is no necessary concordance between biological characteristics and culturally defined groups. On every continent, there are diverse populations that differ in language, economy, and culture. There is no national, religious, linguistic or cultural group or economic class that constitutes a race . . . there is no causal linkage between these physical and behavioral traits, and therefore it is not justifiable to attribute cultural characteristics to genetic inheritance.[4]

However, there are others who answer this question with a resounding "yes:"

The three-way pattern of race differences is true for growth rates, life span, personality, family functioning, criminality, and success in social organization. Black babies mature faster than White babies; Oriental babies mature slower than Whites. The same pattern is true for sexual maturity, out of wedlock births, and even child abuse. Around the world, Blacks have the highest crime rate, Orientals the least, Whites fall in between. The same pattern is true for personality. Blacks are the most outgoing and even have the highest self-esteem. Orientals are the most willing to delay gratification. Whites fall in between. Blacks die earliest, Whites next, Orientals last, even when all have good medical care. The three-way racial pattern holds up from cradle to grave. (J. Phillipe Rushton, psychologist)[5]

How can there be two such different answers to Harrison's question? One answer states, in dry academic terms, that the popular concept of biological races is not supported by evidence; the other, in straightforward common language, says that there is a three-way pattern of racial differences. One answer is wrong.

HUMANS ARE DIVIDED INTO BIOLOGICAL RACES, OR ARE THEY?

The myth of human biological races is alive and well in our society. Someone like Phillipe Rushton can make claims about racial patterns, even though they are incorrect, and have some popular success because the categories "black" and "white" make sense to us.[6] He uses simple, common language that resonates with some of the cultural patterns we hear about via the media, in our daily lives, and in some versions of history. Rushton's claims are a mix of popular assumptions presented as if they were biological facts. Nowhere in his book *Race, Evolution and Behavior* does Rushton provide any real data to support his assertion that "Blacks," "Whites," and "Orientals" are true biological groups, but he does selectively draw from social statistics on crime, income, and mortality to make spurious analogies and then leaps to connect these to the different evolutionary histories of human races. On the other hand the AAPA statement on race (as well as a multitude of similar statements, peer-reviewed articles, books, and Web sites) states unequivocally that these types of associations are not supported and that the concept of clear or determinate biological races in humans today is not justifiable given what we know about human evolution and biology.

While most people would not fully agree with Rushton about the implications of racial differences, more than would care to admit it probably do see things in his proposal that seem to fit with common perceptions of human variation in the United States: blacks as more athletic and overly sexual, Asians as more bookish and reserved, and whites seem to fall in between, more or less the average everyman. This is because many people today see the division of humanity into races as part of human nature. It's time to bust this myth.

This myth involves the assumption that we can define a specific set of traits that consistently differentiates each race from the other with limited overlap between members. This position also assumes that differences in innate behavior, intelligence, sports abilities, aggression, lawlessness, health and physiology, sexuality, and leadership ability exist between these presumed real clusters of humans and that the clusters can be described as the Asian, black, and white races.[7] Nearly everyone holding these beliefs would accept that these clusters do overlap in many ways and that interbreeding between them is always possible and not necessarily negative. However, as the journalist Guy Harrison put it so succinctly (and sarcastically), the majority of people regardless of what they might say in public believe to some degree in the natural reality of human races. This "reality" is an assertion that we can test scientifically.

Buying into at least some of this myth about races also suggests a suite of correlates. One is that since these differences are "natural," we should probably be wary of spending much social and economic capital trying to correct them. Some may also feel that the civil rights movement of the last century and the 2008 election of a black American president indicates that US society has already done as much as is possible to ameliorate racial inequality. From this perspective, focusing on race is not really that important anymore. Finally, many might argue that if race is not a biological entity, then how can the actual, and well-documented, differences in health, sports participation, test scores, and economic achievement between the "races" in the United States be explained? In the same vein, what about ancestry tests? How can a company test our DNA and tell us that we are 40 percent Kenyan or 60 percent Irish? Isn't that about race?

Testing Core Assumptions about Race

To bust the myth of race we have to test the core assumptions and refute them.

ASSUMPTION: *Human races are biological units.*
TEST: Is there a set of biological characteristics that naturally divide up humans beings into races? If yes, then the assumption is supported; if no, then it is refuted.

ASSUMPTION: *We live in a (mostly) postracial society.*
TEST: Does our society still use race in assessment, definitions, and daily life? If no, then the assumption is supported; if yes, then it is refuted.

ASSUMPTION: *If race is not a biological category, then racism is not that powerful or important in shaping human lives.*
TEST: Can we demonstrate that racism, without the existence of biological races, is a significant factor affecting human health, well-being, and access to societal goods? If yes, then the assumption is refuted; if no, then it is supported.

ASSUMPTION: *If we can see consistent differences in sports, disease patterns, and other areas tied to physical features between races, these must reflect innate differences between these groups of people.*
TEST: Are these differences consistent over time? Are they due to biological or unique racial characteristics or are they better attributed to other causes? If yes, and they can be linked to biological patterns of human groups, then the assumption is supported; if no, then it is refuted.

If we can refute all four assumptions, the myth is busted.

MYTH BUSTING: RACE ≠ BIOLOGICAL GROUPS

Although humans vary biologically, we can demonstrate that this variation does not cluster into racial groups. What we refer to as human races are not biological units. Many articles, books, and official statements make this point. However, there are very few brief and succinct overviews of human biological diversity as it relates to racial typologies. Reviewing information about blood groups, genetics, and morphological and physiological variation in the context of evolutionary processes demonstrates unequivocally that there is no way to divide humanity into biological units that correspond to the categories black, white, or Asian, or any other categories.

For close to three hundred years people have been trying to name and classify racial grouping of humans. Carolus Linnaeus, the father of modern taxonomy, made the most important attempt to do so and his classifications still seem very much like current ones.[8] Linnaeus saw the distinction among groups of humans as being rooted in their continental origins (Africa, Asia, Europe, Americas). He saw all humans as belonging to one species, *Homo sapiens,* with a number of subspecies representing the different races.[9] In the tenth edition of his major taxonomy of everything, *Systema Naturae,* published in 1758, Linnaeus proposed four subspecies (races) of *Homo sapiens:* americanus, asiaticus, africanus, and europeanus (he added a fifth category, monstrosous, as a catch-all for wild men and mythical beasts). Unlike his other classifications, which were based on drawings and anatomical analyses of specimens, Linnaeus based his division of humans on what he heard and read about the peoples of the different continents.

> *Homo sapiens americanus* was "red, ill-tempered, subjugated. Hair black, straight, thick; Nostrils wide; Face harsh, Beard scanty. Obstinate, contented, free. Paints himself with red lines. Ruled by custom." *Homo sapiens europeaus* was "white, serious, strong. Hair blond, flowing. Eyes blue. Active, very smart, inventive. Covered by tight clothing. Ruled by laws." *Homo sapiens asiaticus* was "yellow, melancholy, greedy. Hair black. Eyes dark. Severe, haughty, desirous. Covered by loose garments. Ruled by opinion." And last (and obviously least) *Homo sapiens africanus:* "black, impassive, lazy. Hair kinked. Skin silky. Nose flat. Lips thick. Women with genital flap; breasts large. Crafty, slow, foolish. Anoints himself with grease. Ruled by caprice."[10]

These descriptions initiated the still common mistake of mixing presumed cultural differences with biological realities. The anthropologist Jon Marks has repeatedly pointed out that if you read them carefully, Linnaeus's race descriptions sound a lot like those of Rushton's and other modern racialists.

About half a century after Linnaeus the German naturalist Johann Friedrich Blumenbach developed another set of nonscientific human racial classifications, based on geographical definitions and some facets of skull morphology. His classifications included Caucasian, Mongolian, Malayan, American, and Negroid races, which were also referred to as white, yellow, brown, red, and black (based on serious ignorance about skin colors around the planet). Finally, during the mid-twentieth century the physical anthropologist Carleton Coon developed a derivation of Blumenbach's races with a more refined set of skull measurements that is still used by some racial topologists today: the Capoid race (southern and eastern Africa), Caucasian race (western and northern Europeans), Mongoloid race (Asian and Americans), Negroid (or Congoid) race (all of Africa aside from parts to the south and east), and the Australoid race (Australians). Most importantly Coon proposed that each of these races had a separate evolutionary history and thus a suite of behavioral and other traits that evolved separately.[11]

Despite attempts by researchers over the centuries to divide humans into races based on skull shape, geographic location, and presumed cultural differences, there is absolutely no support for any of these classifications (neither those mentioned above nor the countless others proposed) as actually reflecting the ways in which the human skull, genetic characteristics, or other phenotypes cluster in our species.[12] So what does human biological variation actually look like?

As pointed out in our discussion of evolution and genetics in chapter 3, we look at variation in populations. Populations are collections of people that reside in more or less the same place, or in different places but are constantly connected, and mate more with one another than with members of other populations. There are thousands of populations of the species *Homo sapiens* spread across the globe. And in some areas (large international cities like New York, London, or Singapore) individuals from many of those populations congregate. To define a race, then, we need to be able to identify a population or set of populations that has a suite of unique markers that differentiate it from all other such populations and mark it as being affected by slightly different evolutionary forces so as to have altered genetic patterns relative to the rest of the species. Let's look at how we vary biologically between and within populations in our blood, immune system, genetics, body shape and size, skin color, and skull shape.

Blood

For centuries people have looked at blood to tell us about humanity. We know that blood is important (lose enough and you die) and during the last century researchers began to discover that blood itself is made up of a number of different elements, all of which vary a bit. Basically, blood is made up primarily of red blood cells (for oxygen transport), white blood cells (defense against infection), platelets (for clotting), and plasma (the liquid part of blood). There are also a number of other things associated with these main components and even others that use the circulatory system to get to different parts of the body.[13]

Many sets of proteins serve a variety of functions associated with red blood cells. We call these protein sets blood types.[14] The best-known blood type classification is the ABO system, which is often coupled with another system, the Rhesus blood type, noted as positive (Rh+) or negative (Rh-). Today we can track more than fifteen blood type systems whose alleles (forms of genes) are found in variable frequencies across different human populations.

In the ABO gene there are four alleles: A1, A2, B, and O. A1 and A2 are very similar, and mostly respond identically. The three main alleles, A, B, and O, have a set of relationships with one another, in which A and B are considered dominant to O and codominant to one another.[15] In other words, the eventual phenotype of the genotypes AA and AO is A; that of BB and BO is B; that of OO is O; and that of AB is AB. Across the human species these alleles are found at the following frequencies: 62.5 percent O, 21.5 percent A, and 16 percent B. But if we look at the level of different human populations we see different distributions of these alleles. For example, the frequency of allele B is at, or nearly at, zero in many indigenous populations in South America, southern Africa, northern Siberia, and Australia, and higher than 16 percent in indigenous populations in central Asia (figure 2), central West Africa, northern Russia, and mainland Southeast Asia.[16] Alternatively, the A allele is found at its highest frequencies (more than 40 percent) in the Saami (an indigenous population) of northernmost Europe and in some groups of Australian Aborigines.[17] Are populations that share these similar frequencies of A or B more closely related to one another than to the populations next to them that have different frequencies? No.

Understanding natural selection and gene flow helps us understand the distributions of blood types. Probably the most common allele is O because it is the original allele, while A and B are more recent mutations identical to O but with the addition of an extra sugar group. Also, the different ABO phenotypes confer different slightly different support against diseases. Specific blood types may increase or decrease

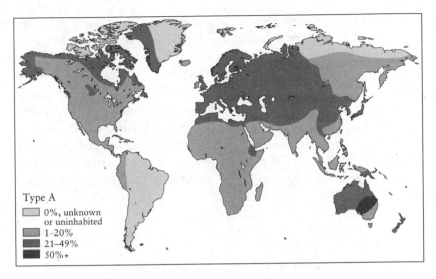

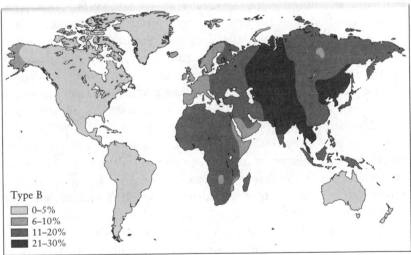

FIGURE 2. Geographical distribution and frequencies of the blood types A and B. Note that they do not follow the big three racial division of European, African, and Asian. Adapted from A. Fuentes (2011), *Biological Anthropology: Concepts and Connections*, 2nd ed. (Burr Ridge, IL: McGraw Hill Higher Education).

chances of surviving things like malaria or other blood-based parasites. However, the majority of variation in blood groups comes from the movements of human populations over the past 50,000 years or so. Gene flow is the major evolutionary force acting on distribution of the ABO alleles across human populations. None of these alleles are unique to specific populations, nor are their frequencies. And most importantly, none of the patterns of ABO (or other blood groups) match up with the black-white-Asian model of dividing humans into racial categories. In fact, the full range of blood variation is found in nearly every single human population. The biology of blood does not support biological race.

Immune System and Disease

Natural selection, gene flow, genetic drift, and mutation have combined with complex cultural patterns to make disease a major factor in recent human evolution. This became especially important as humans began living in towns and villages alongside their farming and domesticated animals. Today humans are more spread across the planet than any other mammal. We live in more types of places (mountains, cities, rainforests, deserts, etc.) and as a result, encounter a larger variety of things that can cause disease. So the immune system is an important part of human evolution.

A major part of our immune system is the human leukocyte antigen (HLA) system. Made up of a series of proteins on the surface of white blood cells, the HLA system recognizes potential infectious agents (things that are foreign to our bodies). These HLA proteins are able to tell the difference between self (your own body) and other (foreign proteins and pathogens) because of specific chemical structures. These proteins then signal the immune system that foreign substances are in the body, so that our bodies can mount a defense.

The HLA system is one of the most variable genetic systems in humans, with between five and seven genes involved, each having from three to more than a hundred alleles. This means the HLA system in people displays an enormous array of genetic combinations. Very few people, even within the same family, will share the exact same system. This means that in any human population there is a lot of variation in immune system response to pathogens. Because humans have spread around the globe so extensively in the past 50,000 years, natural selection is partially responsible for this variability. Having high diversity in the HLA system within populations increased chances of individuals surviving and reproducing because greater variation among our immune systems betters the chances that at least some individuals within a population will be able to combat new pathogens.[18]

Other diseases, not related to HLA function, also have their origins in the variation in allele frequency across human populations. Albinism, the lack of production of pigment in skin, comes from a set of rare alleles for genes on three different chromosomes. Albinism comes in two types. One type emerges from a set of alleles that result in restriction of the production of the enzyme tyrosinase (required for pigment production). This form of albinism is primarily in populations from parts of Europe and central Asia. In the second type of albinism tyrosinase is produced but a failure occurs later in the production of pigment. This second type is most common in parts of Africa and the Americas. Interestingly, because each type has a functioning pigment system except for one component, two individuals with the two types of albinism can mate and produce a nonalbino child, because their allele patterns complement one another.[19] Many other genetic disorders occur more commonly in some human populations than others owing to mutation, gene flow (or lack thereof), genetic drift, and cultural and ecological or environmental factors. One of the best known, and most often racialized, examples is sickle cell disease. This disease is often associated with peoples of western African descent and is held up as an example supporting a "black" race. Turns out, this is not at all the case. Sickle cell disease is a blood disorder that can occur in individuals who carry two copies of a specific allele for a protein that is part of the oxygen-transporting system in red blood cells. In times of stress, like malnutrition, exhaustion, and other diseases, the protein causes some red blood cells to become sickle-shaped, preventing effective oxygen transport. The result is an illness that severely weakens the individual and puts them at risk for other diseases.

The mutation that causes the sickling allele shows up in many human populations at very low levels (there are at least five independent mutations that have this same effect). But in most cases the mutation tends to disappear quickly, or stay at extremely low levels, because of its negative impacts. However, these alleles can be found in relatively high frequencies in some populations in western Africa, the Arabian Peninsula, southern India, and Central America. Note that only one of these geographic areas corresponds to the race category "black." What does unite these widely separated geographic areas is the presence of the powerful disease malaria.

Malaria is caused by a group of parasites that spend part of their life cycle in mosquitoes, and if an infected mosquito takes blood from another organism (that is, bites a person), the parasite can be transferred. In people the parasite can cause problems in the circulatory and respiratory systems, sometime resulting in death. Malaria is a problem for humans only if a lot of mosquitoes are around. Interestingly, individuals with sickle cell disease (those that have two copies of the recessive sickling allele) generally do not contract malaria. Even a little sickling of red blood cells inhibits the reproduction of the malaria parasite. Individuals who have one sickling and one regular allele get mild sickling but do not get the full-blown disease; these individuals have some immunity to malaria. It appears that in areas with a high risk of malaria, the mutant allele (one of the five) can remain in the population at relatively high frequencies. These areas appear to be where people for thousands of years have cleared forested areas for agriculture, creating open fields and many places for stagnant water, which in turn attracts mosquitoes. Human alteration of the environment (niche construction) changes selection pressures on humans, mosquitoes, and the malaria parasite. The result is higher-than-expected frequencies of the sickle cell allele in certain human populations.

Interestingly this process did not happen everywhere that malaria occurs. In some areas humans changed the environment too recently for evolutionary changes to occur and in others chance plays a role. Mutation is fairly random and an effective mutation has to co-occur with the appropriate conditions for allele frequencies to change significantly. There is also a modern biocultural part to this story. As humans move across the planet, they change allele frequencies via gene flow. For example, migrations from the Arabian Peninsula, India, and western Africa resulted in higher frequencies of the sickle cell alleles in North America. Again, we see that human cultural behavior such as migration influences evolution and variation, this time via gene flow.[20] However, even in the United States where sickle cell is thought of as a "black" disease it is found in many individuals who are not black. Sickle cell disease does not support racial categories.

Humans vary in their immune and disease systems because of migration, gene flow, cultural shifts (towns, domestication, etc.), and contact with a wide array of environments over the last few hundred thousand years. The majority of variation in our species is found in almost every living population, and neither HLA nor disease patterns match the black-white-Asian categories.

Genetic Variation

The Human Genome Project, completed in 1998, was designed to identify all the genetic material in humans. The very first draft of the project confirmed what many anthropologists, biologists, and geneticists had been saying for nearly fifty years: humans, as a species, demonstrate little genetic variation between populations. In 1972 the biologist Richard Lewontin pointed out that despite the wide variability in human DNA sequences, the majority of our genetic variation can be found in every living population. Since then more extensive research has confirmed that despite nearly seven billion *Homo sapiens* spread across the planet and our enormous range of body size, shape, color, and form, the vast majority of human genetic variation is found *within* populations rather than between populations.[21] In other words, all human populations—Japanese or Swedish, Australian Aborigine or African, American Indian or Russian, and so on—share extremely similar genetic makeup.[22]

The real confusion between human genetic variation and the race concept comes from the fact that while our overall genetic makeup can be almost identical across human populations (at the macro level) it is at the same time really quite diverse on the micro level. In fact, if we look at very small areas of the genome we can identify genetic variants that are more common in certain areas of the globe than others. To an extent, we can even attempt to identify the genetic histories of individuals, a snapshot of the lines of people in the past who have contributed to one's genetic ancestry, by examining these frequencies of patterns in the micro-level variants of individuals' DNA. How can we be so similar and yet have this micro-level diversity at the same time? Basic genetic analyses can help us understand this pattern of similarity and diversity and show us that it does not equate with any of the categories of human race.

A common way to assess genetic variation is to look at how much variation is found between populations as opposed to within populations. Comparing any two populations, theoretically they can range anywhere from 0.0 (identical for every genetic variation and frequency of those variants) to 1.0 (different for every variation and frequency). Multiple researchers and research groups have looked at thousands of human genes and multiple other stretches of DNA and found that (when there are differences), most scores range from .03 to .24 (averaging about .16 or .17). At few specific spots on the DNA, values are as high as .4 or more, but these are extremely rare.[23]

What this means is that across the human genome, the vast majority of genetic variation is found within populations, and relatively little is found between populations (remember, we are talking about populations, we have not even gotten to races yet). This is an amazingly low interpopulation variation for mammals, especially large-bodied mammals that can move over great distances. For example, multiple white-tailed deer populations found across a few states in the southeastern United States have an average differentiation of about .7. In other words, we find more genetic variation between a population of deer from northern North Carolina compared with one from Florida than we do between human populations from Central America, central Asia, and central Africa. Even more to the point, if you compare any two people from anywhere on the planet and then any two chimpanzees, the chimpanzees would have 75 percent more differences with each other than would the people.[24] None of the examined variations map onto the traditional race categories. There were no genetic patterns that identify and lump whites versus blacks versus Asians; these patterns were looked for extensively and found not to be present.

The recent white paper study by the American Society for Human Genetics states that "because different parts of the genome have different ancestral histories, different marker systems often provide somewhat different information about population history and individual ancestry." However, the same paper also asserts that "the routine treatment, in science, of ancestral, ethnic, and so-called racial groups as bounded biological entities perpetuates an inaccurate concept of human variation and increases the possibility of stigmatization and discrimination of the groups and the people within them on the basis of traits, behaviors, diseases, and other attributes."[25] Basically, this means that we can use some genetic variants to describe patterns within and between populations, but these are not races nor do they divide humans into racial types or categories.

What does this mean? We know that human populations have more, genetically, in common with one another than they differ. At the same time there are specific aspects of the genome, called ancestry informative markers (AIMs) that can help identify the population histories of individuals. Remember, all individuals have parents, who had parents, who had parents, and so on; thus a population history is the tracking backward of these ancestors and seeing if they are identifiable with specific patterns of genetic variation. These AIMs are generally single nucleotide polymorphisms (the smallest identifiable segments of the DNA) that occur with high frequencies in particular populations or population clusters.[26] They do not lump into categories such as "white" or "black," but they do show up in particular frequencies in populations associated with specific geographic and cultural clusters we call Yoruba, Finnish, Japanese, Saami, and so on. We have a large (but incomplete) database of many populations from around the globe and we can identify small patterns of genetic diversity in these populations that differ from patterns in other populations. Not all individuals in a population have the same pattern of genetic variants, but many (or even most) do. This is similar to the situations with malaria and sickle cell disease, or the blood group patterns and other genetic traits, in the sense that some biological patterns can cluster in different areas of the globe. However, AIMs are not usually functional aspects of the genome (as in the sickle cell example) but rather are elements structured by gene flow.[27] So, populations that mate more within their groups than without would be expected to share certain micro-level patterns of DNA more in common with one another than with other such populations. AIMs and the whole process of ancestry testing can help us gain insight into our genetic histories, but the data and analyses do not provide any support for the existence of races and the reference samples used in these studies remain limited.[28]

In all of this scrutiny of human genetic variation there is an extremely important finding about the African continent that affects our understandings of genetics and race: there is nearly twice as much genetic variation among human populations in Africa than among all populations outside the African continent. There also is more DNA sequence variation within Africa than outside Africa.[29] That is, all the genetic variation in the world is a subset of the variation found in populations on the African continent. This is because modern humans have been in Africa longer than anywhere else on the planet.[30] Variation needs time to accumulate, thus the areas with the highest degree of variation will be the areas where humans have resided the longest. So as figure 3 shows, species-wide human genetic variation does not support the concept of three overlapping races; rather, it demonstrates a single human race and the fact that we have a lot of gene flow and a recent shared ancestry in Africa. The patterns in our DNA do not support the concept of discrete races in humans.

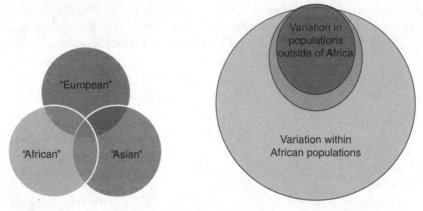

FIGURE 3. Most people think of genetic diversity in our species as three overlapping circles, but the reality is shown here. There is more genetic variation in Africa than everywhere else combined. Venn diagram of human genetic diversity based on data from N. Yu et al. (2002), Greater genetic differences within Africans than between Africans and Eurasians, *Genetics* 161: 269–274. Adapted from diagram by Jeffrey C. Long. Reproduced by permission of American Anthropological Association.

Variation in Body Shape and Size

Humans vary substantially in the size and shape of their bodies. This variation shows up both in the relative contributions of particular body parts (legs, arms, head) to overall body shape and in overall height and the shape of the torso. There is also a lot of variation in the pattern and density of body fat and muscle. The body mass index (BMI), a measurement of weight relative to height, is often used to assess patterned variations in human size and shape.

Average body mass (measured as weight within each sex) varies by as much as 50 percent across human populations, meaning that the largest humans are half again as heavy as the smallest (or more in some extreme cases). BMIs (as population averages) range from about 17 to 25 across human populations. The width of the human body at the pelvis varies by about 25 percent across our species, and average heights range from about just under five feet to about six feet (from about 150 to 185 cm). Aside from the extreme ends of the height spectrum, the human species exhibits about a 10 percent variation in height overall. Sexual dimorphism, or differences in size between males and females, is about 15 percent, meaning that male bodies, on average, are larger than female bodies. Interestingly it looks as though as early human societies transitioned to food growing, and as global climates warmed up in the past 10,000 years or so, human bodies decreased in body mass. More recently, in many developed nations, people have shown some pronounced increases in height and body mass, as health care and nutritional patterns improve dramatically.[31]

Why is there such extensive variation in body shape and size in humans? It turns out that work by zoologists John Allen and Carl Bergmann on mammalian body form helps us understand human shapes. They found that in environments stressed by cold (the Arctic, for example), mammals tend to have increased body mass relative to body surface area (think of penguins or seals, with their stocky and relatively squat bodies). In environments where mammals are stressed by heat (deserts and tropical savannahs, for example), the opposite is true: mammals have a decreased body mass and increased body surface area (for example, giraffes or the ears on elephants). Mammals can live only within a relatively small range of body temperatures; they must constantly retain or lose heat when they are in environments that are above or below those temperatures. So body size and shape are influenced by natural selection, as those variants that do better under thermal stresses become most common in a population.

As mammals, humans display this same pattern of morphological variation. Human populations that have spent many generations in cold-stress environments have larger torsos and shorter, stockier arms relative to many other human populations. For example, think of groups from the extreme north, like Alaskan natives and the Saami peoples of northwestern Eurasia. These body proportions maximize mass and minimize surface area, resulting in more efficient heat retention. Of course, humans also adapt to cold environments with cultural adaptations such as clothing and fire, by means of niche construction. The reverse is true of some populations that have lived for long periods in heat-stress environments. Here we see either very tall bodies with long arms and legs or very small bodies with proportional limbs. In both of these cases, surface area is maximized and mass minimized, increasing the effectiveness of heat loss.

Although natural selection has clearly influenced human body form in climactically stressful environments, the majority of humans do not live in such environments, so most populations are not under this selective pressure. So how to account for all the variation in body size and shape we see in humans? Think of the interaction between genetics, development, and selection as setting the range of possible shapes for humans, and then note the effects of cultural and nutritional factors.

Because most populations are not under strong climactic stress, the selection pressures on body size and shape are fairly relaxed, and a wide array of variation can be expressed.

Human body variation today is largely shaped by gene flow, mating patterns, and nutrition. In a population with little gene flow with other populations and fairly equally distributed nutrition, individuals tend to converge at a similar body shape and size, whereas in populations with high rates of gene flow and variable nutrition, individuals are more variable. Because of flexibility in responses to environmental stresses, each human population has a good deal of genetic and developmental variation underlying potential ranges in body size and shape. As in the case of sickle cell disease, processes of evolution (natural selection, gene flow) are integrated with cultural patterns (migration, mating, and material and nutritional culture) to influence the shapes and sizes of human bodies. In general, relative geographic region (extreme north or south) correlates somewhat with body mass and width. Peoples in regions closer to the extreme north (such as northern Europe) and south (such as southern Chile) of the planet (with more extreme climates) tend to be large bodied. These correlations do not hold for height, however. Height tends to vary less within populations and more between populations. However, migrations, dietary customs, activity patterns, diseases, and, of course, the parameters maintained by natural selection affect the size and shape of our bodies. Most importantly, tall, short, thin, heavy, high BMI, and low BMI populations do not map to the three racial categories, and in each of the areas associated with the big three races (Eurasia, Asia, and Africa) populations of nearly all body types and shapes can be found. The diversity of human body size and shape does not support the division in humanity into white-black-Asian racial categories.

Human Skin Color

The most overemphasized and misunderstood aspect of human variation is skin color. Although many people think that skin color is a good biological way to classify people and that it identifies race, this belief is incorrect.[32]

The differences in human skin are not really about color at all. Human skin has only one main pigment—melanin, which only comes in the colors of black and brown. In addition to melanin, the thickness of the skin, the blood vessels (and the blood in them), and a minor pigment called carotene (orange-yellow) also have minor roles in skin coloration. What makes a difference in variation in skin coloration is the distribution and production of melanin and a few related biological components in the skin, which together result in varying intensities of light absorption and reflectance, making skin look darker or lighter.

Melanin is produced between layers of the skin (the dermis and the epidermis). The dermis (the inner layers) has the blood vessels, hair follicles, and glands (largely sweat glands). The epidermis (the outer layers) is primarily cells that continuously divide and replace themselves, moving toward the outermost layers; these outer layers are what we generally think of as skin. In between the dermis and the epidermis lives a type of cell called a melanocyte. Melanocytes produce melanin and distribute it into the cells of the epidermis. As the epidermal cells divide and move into the outer layers, they bring the melanin with them and distribute it across the epidermis. The density and distribution of melanin cause different levels of reflection and absorption of light in the skin and thus the appearance of different skin colors.

The number of melanocytes does not vary significantly from one human to another, but the density of melanin does. The more melanin that is produced and distributed to the epidermis, the less one type of light (white light) is reflected and the more another type of light (ultraviolet, or UV, light) is blocked from entering the dermis. So, if a person's melanocytes are producing large amounts of melanin that is being effectively distributed throughout the epidermis, that individual will look darker (reflect less light) than an individual with less active melanin production and distribution. Because individuals with less reflection have more melanin in the epidermis, their skin can prevent more UV light from reaching the dermis.

The baseline variation in human skin color arises from a specific kind of environmental pressure. Ultraviolet light in high doses can cause severe damage to layers of the dermis and even plays a role in initiating skin cancer and disrupting other aspects of physiological functioning. This is why doctors recommend using strong sun block before spending time at the beach. Until recently, when a hole formed in the ozone layer (the part of the atmosphere that filters UV light) over Antarctica, that ozone layer provided much of the planet with moderate protection from UV light. However, the intensity of UV light has always been greater at lower latitudes (closer to the equator) and less at higher latitudes (nearer the poles). Therefore, natural selection has favored increased rates of melanin production and distribution through the epidermis in areas of higher UV stress. A map of indigenous populations around the globe shows exactly this: UV stress tends to correlate with darker skin color. Substantial research supports the hypothesis that this relationship between melanin density and UV light is the basis of variation in human skin reflectance.[33]

Why aren't all humans dark? Humans do need a small amount of UV light to penetrate into the dermis. In low levels, UV light assists in the production of vitamin D, which is important for healthy skin, bones, and metabolism. Human populations near the higher latitudes (either far north or south), where UV intensity is lower, face the potential problem of not getting enough UV light for sufficient vitamin D production. These conditions would favor less intense production and distribution of melanin. Again, this pattern can be seen around the planet, with darker skins clustering toward the equatorial regions and lighter skins found further north and south. In short, UV light intensity in the environment has affected human populations, and the resultant adaptation (relative melanin production/density) helps explain the variation in human skin reflectance levels.

But variation in human skin color is more complex than just melanin distribution in populations living near or far from the equator. For example, all humans have a limited ability to respond to increased stress from UV light through tanning. When we tan, our melanocytes temporarily increase their melanin output in response to UV exposure. Melanocyte function, like other functions of our bodies, also varies in effectiveness with age, health, and a variety of diseases. Finally, movement by humans both far north and far south of the equator and gene flow between populations have resulted in a mixing of the adaptations to UV light with other factors. Thus, while natural selection sets the range of current skin color, what people look like in any given population is modified and distributed by gene flow and cultural patterns such as the use of clothing and artificial or natural tanning.

Although skin color varies across the human species, latitude accounts for most of the variation; very little variation occurs among populations within one large region or within a population (figure 4). One can find darker-skinned populations in lower latitudes, including sub-Saharan Africa, south Asia, Southeast Asia, and Polynesia. Lighter-skinned populations are found in northern latitudes, including the Americas, northeast Asia, and northern Eurasia (Europe). The main exception to this pattern is found in populations, such as in the United States or Brazil, that have

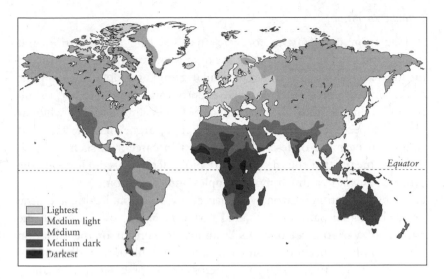

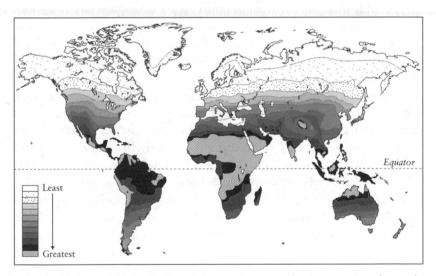

FIGURE 4. Geographical distribution of skin color patterns (top) and UV light (bottom). Note how the "darkness" of skin color maps very well to the incidence of UV light. Adapted from A. Fuentes (2011), *Biological Anthropology: Concepts and Connections*, 2nd ed. (Burr Ridge, IL: McGraw Hill Higher Education).

experienced large, recent migrations from various regions around the planet. We cannot use skin color to characterize specific populations. We can say only that it varies by regions of lower or higher latitudes. The current patterns of pigment distribution and skin color are due largely to human adaptation and gene flow and do not support the division of humans into black, white and Asian races.

Cranial Variation

A long-term mainstay in forensic analyses (the identification and description of dead people) is the use of cranial (skull) measurements to identify skeletal remains with a particular group of people. For example, in the United States today, forensic experts can usually classify a skull from the United States into the categories of Asian, black, or white at about 80 percent accuracy. Some researchers argue that the fact that skulls can be more or less reliably placed in categories such as black or white means that the categories are biologically based. However, the evidence suggests a much more complex interpretation.

Cranial variation is strongly influenced by nutrition, health, and gene flow. The actual patterns of cranial variation in our species match the pattern of variation in our DNA; about 80 percent of the variation in cranial shape occurs within each human population, and about 20 percent occurs between populations across geographic regions.[34] So, how is it possible that forensic scientists can classify skulls into race categories here in the United States with such relatively high levels of accuracy? The answer lies in the ways the crania are classified. If, for example, we have three categories in which to place a set of crania, we can only cluster them into those three. However, if we had six or eight or ten categories, we could cluster them that way as well. For example, in recent work forensic scientists easily differentiated the crania of white American males dating from 1979 from the crania of white American males dating from 1840, and it was as easy to do this as to classify the crania of modern American white and black males from each other.[35] Does that mean that white American males living in 1979 belong to a different race from white American males living in 1840? No, of course not. Numerous studies have shown that cranial form changes measurably across time within a population. If subgroups within populations or regions differ in health, nutrition, and gene flow then some measurable cranial differences will show up, especially if these subgroups of populations derive some ancestry from diverse geographical regions on the planet. Also, remember that even something as simple as limited gene flow can make two populations look more or less alike—we do not need to invoke race concepts.

It is also extremely important to note that skull measurements of humans in the United States would not be exactly the same as skull measurements taken in other parts of the world, given the differences in populations and morphologies. In other words, measurements indicating "black" in the United States would not even come close to classifying all crania from populations on the African continent (especially because there is more genetic diversity between African populations than all populations outside of Africa). The same holds true of measurements indicating "Asian" in the United States and any attempt to encompass the diversity of skull morphology found in Asia (which has two-thirds of the world's population). In addition, none of the actual cranial measurements or patterns used to identify groups is unique to any of the big three race categories. The divisions are based on averages and ranges, so any specific cranium may or may not fit within the "correct" range. This is why experts do make a certain number of errors when placing crania in categories. The differences between crania and between groups are those of degree, not of kind.

A great deal of the cranial variation we actually notice has more to do with face shape and form and hair than the actual construction and overall shape of the skull. These characteristics are even less useful for classifying peoples, as types of hair (frizzy, thick, dark, light, etc.) are distributed across the planet in ways that generally do not correlate with specific patterns in skin color, face shape, body type, or geographic origin.[36] The same is true for broad and thin noses, lip size, and the shape and structure of cheekbones and chins. Cranial variation cannot be used to sort human beings into racial categories.

There is no support for biological races

We can look to human biology to understand how people vary, how populations differ from one another, and how patterns of adaptation and gene flow shape the way humans look across the planet. Data and results from research into body shapes and size, genetics, skin color, skull shapes, and every other aspect of human biological variation demonstrate unequivocally that we cannot divide humans into discrete biological clusters of white, black and Asian. This does not mean that humans do not vary—populations do differ from one another and this variation can be important. It just means that the racial divisions white, black and Asian do not reflect biology: they are cultural constructs.

Why don't most people know this? In large part it is because of our limited exposure to what humans actually look like. Most people do not have the opportunity to travel across the world and see a large subset of the nearly seven billion members of our species. Nor do they have much opportunity to read concise and accessible summaries of thousands of research efforts documenting human biological variation. As established in chapter 2, we are who we meet. Our schemata are shaped and our perceptions of reality structured by what we are exposed to. For example, look at the picture in figure 5 before reading the next few sentences. You should immediately be able to tell that the three children and the young man are from different populations of humans. Given our shared schemata and experiences you can probably as easily place them into two presumed races: the kids are dark with large noses and frizzy hair, so probably of African origin, and therefore "black" and the young man is lighter with sharp facial features and dark hair, probably southern European origin and therefore "white." If you agree with this assessment then you are half right (the guy in the picture with wristwatch is me, a long time ago). I am of European origin (my father is from Spain and my mother's parents from Eastern Europe) and so would be classified in the United States as white (or Hispanic/Latino, but that is another issue). However, the three kids are not of African descent. We'd call them black here in the United States based on our cultural interpretation of their skin, hair, and faces, assuming these features reflected African descent, but they do not. These kids are members of the Dani people from West Papua (the Indonesian side of the island of Papua New Guinea). They are about as far away as you can get from African descent. (I share many more allele frequencies in common with some African populations than they do.) Our limited personal knowledge of human variation cripples our ability to really understand how erroneous racial assumptions are.

Despite everything we've just discussed, many Americans assume that because we seem able to determine a person's race by looking at them or because we can test our DNA and get a percentage of Yoruba or Irish ancestry using AIMs, then the concept of race must have some biological validity. This is wrong; very few people have the background knowledge to make accurate statements regarding the extent and patterns of human biological variation.

Consider an analogy. Nearly all human beings currently accept the notion that the earth is round. We accept it despite the fact that the earth appears to us in our daily experience to be flat. Only a few humans (for example, astronauts or people who sail around the world and arrive back at the same place) have personally seen or experienced the earth as round. The rest of us accept the evidence as scientifically valid even though our personal experience contradicts it. A similar situation holds with the concept of race. Most people do not have the opportunity to see the patterned distribution of humanity across the globe. Although most of us in the United States can generally classify the people we see

FIGURE 5. These children are not from Africa; they are West Papuan (the Indonesian side of the island of Papua New Guinea). Photograph from author's fieldwork, 1992.

every day into three to five groups (though not always as easily or reliably as one might think), these groupings might not be valid in other locations. Further, these groupings reflect only a small percentage of the global biological variation in humanity. Thus, as with the shape of the earth, the broader situation is not necessarily obvious from our limited perspectives. If we have the context (broad exposure and the scientific data and understandings reviewed here), we can realize that, although our personal experience and cultural context might seem to show us one thing, the overall pattern of human biological diversity demonstrates something else: that *Homo sapiens* is one species, undivided into races or subspecies. The myth that human races are biological units is busted.

MYTH BUSTING: RACE IS NOT BIOLOGY, BUT IT STILL MATTERS IN OUR SOCIETY

Okay, so if races are not biological units and civil rights has made significant changes in our society over the last fifty years, then race does not matter, right? Wrong.

In 2004, 50 years after *Brown v. Board of Education,* the controversies around "race" and racism are raging as brightly as ever. Whether we are talking about the future of affirmative action in elite universities, or what the next U.S. Census form will look like, or what the achievement rates of white males are versus underrepresented students of color, this conversation is by no means finished. (Yolanda Moses, anthropologist)[37]

The point made by Yolanda Moses is that race matters as a social factor in the United States. The concept of race and how it plays out in our society are core factors in structuring our individual schemata and the maintenance of cultural constructs of, and societal expectations for, human behavior. However, in the first and second decades of the twentieth-first century a chorus of voices has emerged arguing that we are moving toward a postracial society, or at least a society where race is no longer as powerful or important as it was for much of the twentieth century.[38] This view contradicts what Moses and the entire American Anthropological Association posit: that race matters as an important cultural component of our society.[39] Although the reality of race and racism as part of our society is not being debated, the relative importance of race is a strong current issue, as noted in a recent poll by ABC News and the *Washington Post*.[40] More than twice as many American blacks identified racism as a "big problem" than did American whites.

Since the 2008 election of Barack Obama as US president, there has been a steady series of debates about the relative role of race and racism in our society—not just about blacks or whites but also about Hispanic/ Latinos and Asians. The improvements in civil rights and the election of a black president do not demonstrate that we are in a (mostly) postracial society. Being black, white, Asian, Latino, or other means something in the United States, and although these categories are not biological units, they are social constructs that are central to many aspects of our society: race is not biology, but it does matter.

Consider the following question: Why is Barack Obama considered black? He is an individual with one parent born in the United States (who would be considered white) and one parent born in Kenya (who would be considered black). Why, when classifying President Obama, do we call him black or African-American and not white or European-American or even better yet Afro-Euro-American? Well, interestingly, this last label is not an option in our classification system; moreover, because of his skin color, hair type, and the fact that one of his parents is black, Obama cannot be white. In the United States we have governmentally crafted definitions of race as well as broadly accepted social definitions. We also practice a form of hypodescent, the notion that racial identity is denoted by physical inheritance and by "blood" from a racial group. But this works in a particular way: the lower ranking group is what defines the descent. So throughout US history (and up to today) "looking" black makes you black, as does any black parentage (even great-grandparents). According to popular opinion, having even one drop of "black blood" in your gene-alogy makes you black, but having many drops of white blood does not make you white.

Why is this? It is tied to the concept that races are biological units and that some races are better than others; thus biological influence (or contamination) from one race dictates what race you are. This is rooted in misguided notions about genetics and biology, but nonetheless remains, subconsciously, a de facto reality for our society. This is one reason why Barack Obama is considered black and not white.

Another reason has to do with our own government's classification system. The Census Bureau creates and maintains a set of definitions that we use to officially classify people in our society. The official guidelines state that

> The Census Bureau collects race data in accordance with guidelines provided by the U.S. Office of Management and Budget (OMB), and these data are based on self-identification. The race response categories shown on the questionnaire are collapsed into the five minimum race groups identified by the OMB, and the Census Bureau's "Some other race" category. The racial categories included in the following text generally reflect a social definition of race recognized in this country, and not an attempt to define race biologically, anthropologically or genetically. In addition, it is recognized that the categories of the race items include racial and national origin or socio-cultural groups. People may choose to report more than one race to indicate their racial mixture, such as "American Indian" and "White."[41]

Note that there is a specific statement that these are purely social categories and not intending to define race as biological. However, as you will see with the following definitions, this is not totally true. Before the census asks about one's race, it first asks if one is "of Hispanic, Latino, or Spanish origin." These categories are not officially considered racial categories (more on this below). Here are the official definitions of race for the US government:

> Mark the "White" box if this person has origins in any of the original peoples of Europe, the Middle East, or North Africa. This includes people who indicate their race as "White" or report entries such as Irish, German, Italian, Lebanese, Near Easterner, Arab, or Polish.

Mark the "Black, African Am., or Negro" box if this person has origins in any of the Black racial groups of Africa. This includes people who indicate their race as "Black, African American, or Negro," or provide written entries such as African American, Afro-American, Kenyan, Nigerian, or Haitian.

Mark the "American Indian or Alaska Native" box if this person has origins in any of the original peoples of North and South America (including Central America) and who maintain tribal affiliation or community attachment. This category includes people who indicate their race as "American Indian or Alaska Native," and/or provide written entries such as Navajo, Blackfeet, Inupiat, Yupik, Canadian Indian, French American Indian, or Spanish American Indian.

Mark any of the Asian boxes if this person has origins of any of the original peoples of the Far East, Southeast Asia, or the Indian subcontinent including, for example, Cambodia, China, India, Japan, Korea, Malaysia, Pakistan, the Philippine Islands, Thailand, and Vietnam. This includes "Asian Indian," "Chinese," "Filipino," "Korean," "Japanese," "Vietnamese," and "Other Asian."

Mark the "Asian Indian" box if this person indicates their race as "Asian Indian" or identifies themselves as Bengalese, Bharat, Dravidian, East Indian, or Goanese.

Mark the "Chinese" box if this person indicates their race as "Chinese" or identifies themselves as Cantonese, or Chinese American. In some census tabulations, written entries of Taiwanese are included with Chinese while in others they are shown separately.

Mark the "Filipino" box if this person indicates their race as "Filipino" or who reports entries such as Philipino, Philipine, or Filipino American.

Mark the "Japanese" box if this person indicates their race as "Japanese" or who reports entries such as Nipponese or Japanese American.

Mark the "Korean" box if this person indicates their race as "Korean" or who provides a response of Korean American.

Mark the "Vietnamese" box if this person indicates their race as "Vietnamese" or who provides a response of Vietnamese American.

Mark the "Other Asian" box if this person provides a write-in response of an Asian group, such as Bangladeshi, Bhutanese, Burmese, Cambodian, Hmong, Laotian, Indochinese, Indonesian, Iwo Jiman, Madagascar, Malaysian, Maldivian, Nepalese, Okinawan, Pakistani, Singaporean, Sri Lankan, Thai, or Other Asian, not specified.

Mark the "Native Hawaiian" box if this person indicates their race as "Native Hawaiian" or identifies themselves as "Part Hawaiian" or "Hawaiian."

Mark the "Guamanian or Chamorro" box if this person indicates their race as such, including written entries of Chamorro or Guam.

Mark the "Samoan" box if this person indicates their race as "Samoan" or who identifies themselves as American Samoan or Western Samoan.

Mark the "Other Pacific Islander" box if this person provides a write-in response of a Pacific Islander group, such as Carolinian, Chuukese (Trukese), Fijian, Kosraean, Melanesian, Micronesian, Northern Mariana Islander, Palauan, Papua New Guinean, Pohnpeian, Polynesian, Solomon Islander, Tahitian, Tokelauan, Tongan, Yapese, or Other Pacific Islander, not specified.

Mark the "Some other race" box if this person is not included in the "White," "Black or African American," "American Indian or Alaska Native," "Asian," and "Native Hawaiian or Other Pacific Islander" race categories described above. Respondents providing entries such as multiracial, mixed, interracial, or a Hispanic, Latino, or Spanish group (for example, Mexican, Puerto Rican, Cuban, or Spanish) in the "Some other race" write-in space are included in this category.

People who are of two or more races may choose to provide two or more races either by checking two or more race response check boxes, by providing multiple responses, or by some combination of check boxes and other responses.[42]

There are a number of relevant factors to be found in these definitions, but one aspect stands out: "black" is treated differently from all the others. If you look closely at the definitions, you will see that "Black, African Am., or Negro" is the only category where the term "racial groups" is used ("if this person has origins in any of the Black racial groups of Africa"). In all of the other main categories the term "original peoples" is used. This marks the black category as a race, a biologized entity, relative to the other categories. Also, note that it is not just any racial groups, but the "Black racial" groups of Africa. Is there any mention of other types of racial groups in Africa (or anywhere else)? No. There is a clear demarcation of "black" as distinct type of category from the other "original peoples" categories. To be sure, the government does explicitly state that "the racial categories included in the following text generally reflect a social definition of race recognized in this country, and not an attempt to define race biologically, anthropologically or genetically." Yet that is exactly what it is doing, indicating with these categories that in the United States race matters and also that there is a hierarchy of races (one that mimics Phillipe Rushton's analyses). There are no reasons given by the Office of Management and Budget for its use of the terms "racial groups" versus "original peoples," but we can look at the history of naming races from Linnaeus to the modern day to see what is going on here. "Black" is associated with a lower ranking in the hierarchy of races. Race matters. It is worth noting that the US government bureau validates this assertion by stating that it is using the "social definition of race recognized in this country."

Examining the other categories, we also see that these ways of classifying people are clearly nonbiological and in fact emerge largely from events and patterns in US history. The classification of Middle Easterners and Arabs as "white" is certainly left over from a time when the relationship between the United States and the Middle East, especially Muslim countries, was quite different. How many in our society today would define Osama bin Laden, Saddam Hussein, Muammar el-Qaddafi, or anyone from Algeria, Morocco, Iran, or Egypt as "white"? The mandate that to be Native American or American Indian you must hail from the "original peoples of North and South America (including Central America)" and "maintain tribal affiliation or community attachment" stems from the history of treaty signings and manipulation of Indian lands and cultures by the US government. Interestingly, this results in a number of Native Americans without tribal affiliation not being legally classifiable as American Indians. The fact that "Asian" applies to anyone with ancestry in the "Far East, Southeast Asia, or the Indian subcontinent," which is about 70 percent of all humans on the planet and a substantial portion of the overall inhabited landmass, emerges from the limited exposure that the United States has had to the wide range of peoples and populations of Asia. Finally, the "some other race" category is a bit of a catchall (except that you can reinsert Hispanic or Latino as a race at this point) for accounting purposes just in case someone comes up with something else. As part of its normative functioning, the government keeps tabs on the socially defined races (in a very general way) in order to manage the country, which invalidates the assertion that race no longer matters. Race is a core part of the United States.

Let's close this section with a few statistics from the US Department of Labor and the Pew Research Center:[43]

- In tests of housing markets conducted by the US Department of Housing and Urban Development (HUD), black and Hispanic potential renters and buyers are discriminated against (relative to whites) nearly 25 percent of the time.

- Light-skinned immigrants in the United States make more money on average than those with darker complexions, and the chief reason appears to be discrimination.

- Blacks and Hispanics have considerably lower earnings than Asians or whites. In 2009, the median usual weekly earnings of full-time wage and salary workers were $601 for blacks and $541 for Hispanics, compared with $880 for Asians and $757 for whites. The earnings of black men ($621) and Hispanic men ($561) were 65 and 60 percent, respectively, of the earnings of Asian men ($952). The earnings of black women ($582) were 75 percent of the earnings of Asian women ($654), a higher ratio than among black and Asian men. The median earnings for white men and women were 89 and 86 percent of their Asian counterparts in 2009. Median earnings for Hispanic women were $509.

- In 2009, about 90 percent of blacks and Asians (twenty-five years of age and older) in the labor force had received at least a high school diploma, the same proportion as whites. In contrast, about 67 percent of Hispanics had completed high school. Asians were most likely to have graduated from college; 59 percent had a bachelor's degree or higher, compared with 35 percent of whites, 24 percent of blacks, and 16 percent of Hispanics. Although blacks and Hispanics were less likely than whites and Asians to have obtained a college degree, the proportion of college graduates for all groups has increased over time. For all groups, higher levels of education are associated with a greater likelihood of being employed and a lower likelihood of being unemployed. Nonetheless, at nearly every level of education, blacks and Hispanics were more likely to be unemployed in 2009 than Asians or whites.

- The 2008 infant mortality rate per 1,000 births is 5.7 for whites, 13.6 for blacks, 5.6 for Hispanics, and 6.9 for the United States as a whole.

- The 2009 percent of each group living below the poverty level is 11.5 for whites, 32.2 for blacks, 28.4 for Hispanics, 19.4 for other (primarily Asian), and 17.2 for the United States as a whole.

- Percentage of groups without health insurance: 12.2 for whites, 20.9 for blacks, 33.5 for Hispanics, 17.7 for other (primarily Asian), and 17.2 for the United States as a whole.

- The 2009 net worth of US households: white: $113,149, black: $5,677, and Hispanic: $6,325; there is a twentyfold difference between whites and all the others!

I could continue to list statistics, but these are enough to demonstrate the point that, while race is not a biological unit, race as a social reality matters in the United States. The myth that we live mostly in a postracial society and that race does not matter is busted.

MYTH BUSTING: RACE IS NOT BIOLOGY, BUT RACISM AFFECTS BIOLOGY

It is a myth that racism is not a powerful or important force in shaping human lives and that it does not have an impact on human biological systems. Even though race is not a biological unit in humans today, the realities of social race and associated racism and inequality can become biology: race can impact physiological and epidemiological systems. Recent work in anthropology and medicine shows us that inequality and social perceptions of self and other in a racialized society can, and do, have real biological (especially health) impacts, as the sociologist Troy Duster noted in 2005: "There is a complex feedback loop and interaction effect between phenotype and social practices related to that phenotype."[44]

Race is social reality and thus related to patterns of inequality in the United States.[45] People use phenotypic aspects of humans (what we look like) to classify people into races. Thus the reaction by individuals to perceptions of race based on our shared cultural constructs and our schemata can affect the world around us. Actions by others and the way we see ourselves as fitting, or not fitting, into specific parts of society, our expectations of what is normal, and the social niches we occupy can affect the ways in which our bodily systems (our biology) respond to the external environment. This type of impact is especially reflected in aspects of the body that relate to health.

Take the example of hypertension (recurrent high blood pressure). From 1988 to 2006 the overall percentage of US adults with hypertension (age-adjusted) went from 25 to 31 percent. In non-Hispanic white males this increase was the same as the national average, for black males it went from 37 to 42 percent, and in Hispanic (primarily Mexican) males it decreased from 27 to 26 percent. Interestingly, how far someone is living below the poverty level also correlates with a large increase in hypertension over the 1988–2006 time period.[46] Given that both Hispanic and black males earn less on average than whites, and thus have a greater chance of living below the poverty level, one would expect their blood pressures to be equally increased. However, this is not the case; data consistently show that US citizens of African descent (blacks) have higher levels of hypertension than other US residents. Why might this be?

> First, the sociocultural reality of race and racism has biological consequences for racially defined groups. Thus, ironically, biology may provide some of the strongest evidence for the persistence of race and racism as sociocultural phenomena. Second, epidemiological evidence for racial inequalities in health reinforces public understanding of race as biology; this shared understanding, in turn, shapes the questions researchers ask and the ways they interpret their data—reinforcing a racial view of biology. It is a vicious cycle: Social inequalities shape the biology of racialized groups, and embodied inequalities perpetuate a racialized view of human biology. (Clarence Gravlee, anthropologist)[47]

Going on the assumption that "black" is a biological category, the company NitroMed announced that "the African American community is affected at a greater rate by heart failure than that of the corresponding Caucasian population. African Americans between the ages of 45 and 64 are 2.5 times more likely to die from heart failure than Caucasians in the same age range."[48] This was the push they used to get their ethnic/race-specific hypertension drug, BiDil, cleared by the FDA.[49] They developed an antiheart failure drug combination that initially was shown to be ineffective in the general population but had some initial success in a targeted study of black Americans.[50] In addition to the debate about whether or not this drug really does help blacks more than any other group, this is an incorrect response to the problem. It is not the biology of being black that leads to increased hypertension rates. There is no unique or cohesive biological set of characteristics that define "black" or any other race on the planet. Rather, it is the reality of our cultural constructs and the perceptions of race, which result in certain kinds of societal inequality (both perceived and practiced), that affects the health (and thus biology) of people who fall into different race categories. Perceptions and experiences of race affect biology, rather than there being biological differences characterizing the races that lead to the differences observed. Though this is more or less the opposite of what most people think, it is not any less real. This is true for all the racial groupings in the United States, but the case of US blacks and blood pressure is among the most studied, so we focus on that as an example here.

If one takes a look at African descendant populations around the Americas, not just in the United States, we see a very different picture of hypertension. Comparing blacks in the Caribbean with those in the United States we find that many of the Caribbean groups do not show the pattern of disparity in hypertension characteristic of US blacks.[51] This is quite important as there is no biological or genetic suite of characteristics related to blood pressure that differentiate Caribbean blacks from US blacks, but there are social and societal differences. For example, the histories of relationships, colonialism, slavery, and mixing of peoples are somewhat different in the Caribbean than in the United States. Also, a vast majority of Caribbean blacks are considered Hispanics in the US census; however, there is a range of variation in who labels themselves black, who labels themselves Hispanic, or both, or something else entirely. The patterns of inequality and the social construction and definitions of race have different patterns and origins in the Caribbean. In this case, the way in which race is perceived (by oneself and others) in the United States has particular effects on the body.

There are a series of research projects that demonstrate that discrimination, skin color, and the perception of skin color are major factors in the increased hypertension rates in US blacks. Many studies show that there is a correlation between racial inequality and a wide range of physiological measures of stress including increased cardiovascular response, which leads to elevated blood sugar, blood pressure, and heart rate. For example, African American women who were treated unfairly but did not report the discrimination exhibited higher blood pressure than those who spoke up. Also, high status (or wealthier) darker-skinned black men had higher blood pressure than their lighter-skinned counterparts. This is hypothesized to be related to the pattern where the darker-skinned men endured more negative social interactions because darker skin is associated with lower socioeconomic status, according to racial stereotypes in the United States.[52] In an excellent study focusing specifically on this factor the anthropologist Clarence Gravlee and colleagues demonstrated that skin color as a factor in social classification based on culturally defined race categories was a better predictor of blood pressure than a genetic estimate of ancestry (percentage of ancestry from African populations). They also found that using a range of sociocultural variables associated with race as major components of the investigation of hypertension revealed a variety of new pathways to understand the relationships between genetic/biological variability and blood pressure.[53]

In addition to the self-perceptions of race impacting our body's biological systems, the structure of inequality associated with race can also have effects on the health of US populations. For example, an article published in the *New England Journal of Medicine* in 2004 noted that across the United States, black patients generally receive lower-quality health care than white patients. The researchers argued that part of the reason for this might be that black patients receive their care from a "subgroup of physicians whose qualifications or resources are inferior to those of the physicians who treat white patients." The researchers conducted an analysis of 150,391 visits by black Medicare beneficiaries and white Medicare beneficiaries sixty-five years of age or older. They were seen by 4,355 primary care physicians. The study found that the majority of visits by black patients were with a small group of doctors (80 percent of visits were accounted for by 22 percent of physicians), who also provided only a small percentage of care to white patients. Comparing white and black patient visits they found that the doctors visited predominantly by black patients were less likely to be board certified than were the physicians visited by the white patients and also more frequently reported that they were unable to provide high-quality care to all their patients. Finally, the doctors primarily treating black patients reported having greater difficulties obtaining access to high-quality subspecialists, high-quality diagnostic imaging, and non-emergency admission to the hospitals for their patients. The authors of this study concluded that black and white patients are generally treated by different physicians and that the doctors treating black patients may be less well trained and may have less access to important resources than physicians treating white patients.[54] This research shows that social histories and the economic and societal reality of race in the United States can lead to an inequality in access to medical services, which in turn can significantly affect the health (and thus biology) of people classified into different races.

There are a number of other such examples where social perceptions and actions result in inequality that affects our bodies and becomes part of our physiological reality. We can see examples of this across the races and, importantly, we also see this emerging from inequality based on income, education, and social access (all related to poverty and economic hierarchies), and of course, gender. The point being, as discussed in chapter 2, multiple elements go into the construction of our schemata and societal patterns. This is not a myth: these patterns and perceptions of reality are real for us as a culture and in turn have a real effect on the way we see what is normal and natural. This results in societal patterns of inequality that influence and shape our bodies, minds, and health.

Finally, I do want to emphasize that this is not an argument that variation in diseases and disease risk is exclusively determined by social structure and inequality. There is substantial variation across human populations in disease risks and susceptibilities, rooted in genetic variation and patterns of evolutionary pressures. The core point here is that this variation is based on population and population of ancestry and does not map to the racial triad of white, black, and Asian. For example, the deadly Tay Sachs disease is more common in Ashkenazi Jews (Jewish peoples of eastern European origin) than in other ethnic groups. Are Ashkenazi Jews a race? No. Cystic fibrosis is more common in people in the United States with northern European ancestry. Are Northern Europeans a race? No. Sickle cell disease is more common in Americans of African, Arabian, and Indian descent, but do these groups make up a race? No. The point being that there is important variation in allele frequencies across the human species and that variation is distributed within and between populations. This leads to some patterns of disease that affect some populations more than others. But these patterns do not map to racial differences. There are not white, black and Asian diseases. However, being white, black or Asian can put you in different social and environmental realities that lead to inequalities in health and disease. The myth that racism is not a powerful or important force in shaping human biology is busted.

MYTH BUSTING: WHERE DO "RACIAL" DIFFERENCES, OR OUR PERCEPTIONS OF THOSE DIFFERENCES, COME FROM?

Looking around the United States we can see that there are differences in sports, disease patterns, test scores, and many socioeconomic factors among racial groupings in the United States. The myth is that these reflect inherent (natural or biological) differences between these groups of people, but we know by this point in the chapter that this is not true. There is only one human race, and the divisions of white, black and Asian are not biological categories. However, this fact does not alter the reality that there are some patterns of differences, on average, between white, black, and Asian in the United States. The preceding sections demonstrate that while race is not a biological category, it can have effects on health, economic status, or educational attainment, among other factors. Piecing together the information from chapters 1 and 2 we can see that these differences emerge via a complex history and set of social contexts, and the interactions of schemata, history, and political and cultural constructs, but not from innate differences between the groups. Individuals vary, and individuals are found across the spectrum of these differences. The statistics of economic access, health, and education are real, but they are not static. Neither are race categories. Who counts as white, black or Asian changes over time and place, as do the ways in which society reacts to these groups and how we perceive ourselves.

While much of the rest of Lawrence Summer's 2005 speech drew significant criticism (for his assumptions about gender, see chapter 6), the point extracted below is an important one:

> To take a set of diverse examples, the data will, I am confident, reveal that Catholics are substantially underrepresented in investment banking, which is an enormously high-paying profession in our society; that white men are very substantially underrepresented in the National Basketball Association; and that Jews are very substantially underrepresented in farming and in agriculture. These are all phenomena in which one observes underrepresentation, and I think it's important to try to think systematically and clinically about the reasons for underrepresentation. (Lawrence Summers, former president of Harvard University)[55]

We can see differences in representation in sports, professions, socioeconomic status, health, and the like across the races today and we must ask questions about our society in order to explain them. The specific history of the United States, and the perceptions of difference we hold as normal and natural, have structured the ways in which people classified into races live their lives. We have to ask about the waves of migrations to the United States, the history of slavery, and the concept of Manifest Destiny. We need to think about our public education systems, the history of segregation and civil rights, the impact of the Second World War, and the history of our cities and suburbs during the previous two centuries, to lay a baseline for really understanding the social, economic, and health differences we see today.[56]

But these basic and complicated realities are not the first ones many people rely on when thinking about racial differences. Rather, we frequently see characterizations of the races similar to those proposed more than two hundred years ago by Linnaeus and reiterated today by Rushton. Today, generally, most people look around and say that blacks are better at physical sports, whites run companies, and Asians do really well on tests. But are these generalizations really accurate? And if they are, we know that race is not a biological unit, so an explanation for the differences has to be largely nonbiological and thus social and historical.

Let's use the example of racial representation in sports to demonstrate how and why the differences are so much more complex than is evident from a superficial glance. In the 2010–2011 season the National Football League (NFL) had 67 percent black and 31 percent white players (with the other 2 percent made up of Asian, primarily Samoan, and Hispanic/Latino players). Of the senior administrative positions on NFL teams, 16 percent are held by blacks, Hispanics/Latinos, or Asians and 84 percent by whites. Eighty-three percent of quarterbacks, the team leader on the field, are white; 86 percent of the running backs and 84 percent of the wide receivers, both workhorse players for running and scoring, are black.[57] This pattern of the majority of players being black, with even more being running backs and receivers, and the majority of quarterbacks being white in spite of their lower representation in the overall player pool, reflects something. In the National Basketball Association (NBA) as of 2011, 78 percent of players are black, 17 percent white, 4 percent Hispanic/Latino, 1 percent Asian, and 1 percent other (also, 17 percent are not US citizens, most of whom would be classified as Asian and white and possibly Hispanic in the United States).[58] This pattern reflects something. Finally, it turns out that in major league baseball (MLB) two-thirds of all shortstops are Hispanic/Latino (and a majority of those are from the tiny Dominican Republic) and more than one in three MLB players are not from the United States.[59] This pattern also reflects something. But what?

If these patterns reflect a biological reality that whites are better leaders and managers and blacks are better at running and jumping, then white and black must be biological units. This logic would also lead us to the conundrum of trying to find the gene for shortstop in the Dominican population and an explanation for why Asians are so biologically bad at American sports. But these searches would be useless.

These are not group biological differences. There is no evidence that if you randomly select three men, one black, one white, and one Asian, in the United States that the black will be better at fullback or basketball, that the white will be a great quarterback, and that the Asian will not play sports. However, depending on where you select your young men from (city, suburb, rural town, West Coast, East Coast, Southeast, Midwest, etc.), what age you select them at, what economic group you select them from, and their number of siblings, religion, health history, and place of birth of their parents, you are going to have varied results. The point is that the patterns we see today in professional sports are due to historical and social realities, residence patterns, socioeconomic access to sports facilities, and popular perceptions of race differences. It was not that long ago that nonwhites were not allowed to play in most professional sports, that there were no black quarterbacks, and that the majority of running backs were white. The role of race in sports is a social, economic, and historical reality that is neither static nor related to genetics. Instead, it is part of the ever-changing social structure of our society.

This is not to say that there are no aspects of human biological variation at play. If you are a small male, you are not going to be very good at football or basketball (with a few amazing exceptions). There is variation in human muscle quality and density, in hand-eye coordination, and in endurance running, but these are not distributed along racial lines, and all can be radically improved via training. And, of course, elite athletes at the professional level are horrible examples to rely on when examining average differences between people. Professional athletes have risen above thousands and thousands of others at every level to attain the extremely few slots at the top level in their fields. They have also focused the majority of their physiological and social development on attaining professional status in sports, and we know that that sustained effort shapes the body and mind. Why individuals chose a specific sport and what kind of response they receive is very important in understanding these differences. In chapter 1 I invoked the scenario of a bunch of high school kids playing a pickup basketball game; given the choice between three kids of the same height and build, but one white, one black, and one Asian, social histories and current perspectives are going to shape who gets chosen first. This is true across our society. We know biological, social, and physiological development are affected by our schemata and cultural context: if you are from an area where the people around you do not value participation in competitive sports, or most people play hockey, or most people play basketball, there will be developmental and social influences that shape the way you respond to and perceive of sports throughout your life.

There are many excellent overviews of the history of US racial groups and sports and how it has changed over time.[60] In his recent book, Guy Harrison points out that, in spite of the often-heard adage "white men can't jump," nearly all of the best track and field high jumpers are white, and few people classified as black have ever won a gold medal in the Olympics for the high jump. Despite the majority of black players in basketball today, before collegiate basketball was integrated in the late 1950s and 1960s, Jewish athletes (classified as white now, but viewed a bit differently then) dominated the sport, and the majority of heavyweight boxers in the first third of the twentieth century were Jewish and Irish. In the Winter Olympics throughout its history, white athletes have dominated, winning nearly all the medals, with very few blacks competing at all. These patterns in professional sports all have to do with a complicated mix of history, society, and individual variation, and do not derive from any racial ability to be better at sports, math, or management. Regardless of this awareness, the perception of race being associated with innate or natural differences in sports ability is extremely strong in the United States. While this perception is the result of many different histories, there is one that stands out as worthy of our focus.

One Small Piece of Important History: Eugenics

Although it has no biological validity, racial categorization in the United States remains a deeply engrained cultural pattern with potentially negative biological and social impact. A major component of this cultural pattern's resilience in the face of evidence can be traced to the US love affair with an early twentieth-century pseudo-science called eugenics.

In the early 1900s, work on simple genetic systems was becoming widely known, and early geneticists and social theorists adopted this work to develop the field of eugenics, which is the study of human beings with the applied goal of improving human biology and biological potential. The argument was for an enlightened scientific approach to make humans stronger, more disease resistant, and more intelligent. Eugenicists believed that we could improve the human species via careful selective mating and the establishment of human pedigrees; they wanted to make sure "good genes" were protected and "bad genes" were kept out. We now know this was a totally incorrect way to think about genetics, but at the time it made some sense.

The eugenicists were heavily influenced by the idea of simple genetic inheritance, which dominated the early understanding of genetics and was easily grafted onto existing notions of human heredity (such as the idea that you simply get one thing from dad and one from mom and the dominant one is what you have in your phenotype). By the 1920s eugenicists developed a widespread, erroneous conception of simple genetic systems that linked them with stereotypical ethnic traits. For example, eugenicists considered feeblemindedness (low intelligence) to be a simple dominant/recessive trait that they believed occurred with high frequency among immigrants to the United States from southern Europe (Italians, for example, figured predominantly in this categorical disparagement).[61]

Many in the eugenics movement sought to use genetics to explain the social, cultural, and racial differences among groups in the United States. Economic, political, and religious differences were seen as reflecting genetic distinctions. Eugenicists used simplistic ideas about taxonomy and racial categories based on cranial measurement to support their notions. They thought they could predict intelligence, ability to mesh with American society, and a worker's potential from skull size and shape or from skin color and brow size. Their ideas were incorrect, and over time eugenics fell out of favor, especially after World War II, since the Nazis in Germany had used the eugenicist paradigm to bolster their attempts to identify and standardize their ideas about so-called Nordic and Aryan types and to exterminate several groups of people including Jews, Slavs, Romani (Gypsies), and homosexuals.

But the impact of the eugenics movement in the United States remained very powerful. Textbooks in genetics and human biology in use into the 1950s were written by eugenicists, and major arguments against civil rights legislation in the late 1950s and 1960s rested heavily on the published work of eugenicists and their misappropriation of anthropological research. Academically, eugenicists' ideas fell out of favor, but for the public they remain very strong.[62] And this has played a major role in maintaining the myth of innate racial differences.

While we can see real differences in sports participation, disease patterns, and socioeconomic status between the races in the United States, these differences are not due to biological or unique racial characteristics. They arise from individual variation, and social, historical, and economic patterns and contexts that characterize our society. The eugenics movement in the United States played a core role in maintaining the belief that such differences are genetic characterizations of social race groups.[63] The myth of the racial categories black, white and Asian as biological units, or as a natural classification of humanity, remains busted, but many differences between groups in the United States do occur, do change over time, and are a major part of our society and its perceptions of race.

WHAT RACE IS AND WHAT IT IS NOT

The anthropologist Clarence Gravlee has suggested that we stop saying that race is a myth, and instead accept that parts of it are myths while other aspects are not. He is correct: the myth part about race is that in modern humans there are biological races. The nonmyth part is that in our society the social categories of race are a reality that affects our lives. Thus, white, black, and Asian are not real biological, evolutionary, or natural categories nor do they reflect true divisions in human nature. However, white, black and Asian are real categories in the United States, for historical, political, and social reasons. People get placed in these categories both by themselves and by others. These social race divisions have real effects on the bodies and minds of the people in the United States. Race is not biology, but race affects biology, experience, and social context. Here are some closing thoughts on what race is and what it is not.

Race is not a valid way to talk about human biological variation

Biological anthropologists widely agree about how to describe and interpret variation in the human species. This agreement can be summarized in the following five points that represent our core understanding of biological variation in humanity:[64]

1. There is substantial variation among individuals within populations.

2. Some biological variation is divided up between individuals in different populations and also among larger population groupings.

3. Patterns of within-group and between-group variation have been substantially shaped by culture, language, ecology, and geography.

4. Race is not an accurate or productive way to describe human biological variation.

5. Human variation research has important social, biomedical, and forensic implications.

Race is a social reality that can have lasting biological effects

The work of Clarence Gravlee, Bill Dressler, and others discussed in the preceding sections demonstrate this point: race is not biology but it can affect biology. In a February 2000 editorial, the prestigious, peer-reviewed journal *Nature Genetics* issued the following guideline:

> The laudable objective to find means to improve the health conditions for all or for specific populations must not be compromised by the use of race or ethnicity as pseudo-biological variables. From now on, *Nature Genetics* will therefore require that authors explain why they make use of particular ethnic groups or populations, and how classification was achieved. We will ask reviewers to consider these parameters when judging the merits of a manuscript—we hope that this will raise awareness and inspire more rigorous design of genetic and epidemiological studies.

That is, we may use classifications by race and/or ethnicity when talking about human variation, but we must be clear why and how we are using these categories and about issues of directionality and reality of biological groupings. Race as a concept and racial inequality (racism) as a social reality can affect biology.

Race ≠ Ethnicity

Ethnicity is a way of classifying people based on common histories, cultural patterns, social ties, language use, symbolic shared identities, and the like. It lays no claim to biology and is used both by those attempting to classify others and by those within the different ethnic groups as a symbol of social unity. Ethnicity is not a natural set of divisions in humanity; it is fluid, changing over time and space. The terms "ethnicity" and "race" are often used interchangeably, even in commercial ancestry testing; this is wrong. This mistaken usage is a holdover from the patterns established by eugenicists trying to identify as biological groups the various national and ethnic groups who were living in, or entering, the United States in the early twentieth century. From that time on the notion of "ethnic" has been used as a technique for establishing "white" as normal and nonethnic, in contrast to the "other."

Check out the shampoos and hair care products at your neighborhood drugstore: most places will have an aisle or section marked "hair care" and another marked "ethnic products" or "ethnic hair care." This is shorthand for "black," or frizzy, hair care products. Think about the common phrase "ethnic food." Does this refer to what is considered to be typical US (or white) food like hamburgers, hotdogs, or meatloaf? No, it means all the other types of foods associated with nonwhite groups or with subdivisions of southern or eastern European origin, those not considered white in the early parts of the twentieth century, like Jewish, Italian, and Slavic.

The same holds true for commercial ancestry testing. If you submit your DNA sample to one of the many companies that offer such services and your results come back 50 percent Irish, 35 percent German, and 15 percent Yoruba, you might think you were basically "white" but also 15 percent "black." This is a nonsensical statement. The results suggest only that given the limited genetic samples we have to compare your sample with, certain very small parts of your genetic variation seems to fit with the micro-patterns found most commonly in Irish and German samples but there are some small similarities with the patterns found in our Yoruba sample. At best this means that you have mostly western European ancestors, with possible some West African ancestry mixed in. Or the results might be erroneous given the limited sampling of human populations in the reference samples. Irish/German is not equal to "white" and Yoruba is not equal to "black"; they are simply ethnic labels used to refer to the population samples used in the genetic comparisons. This has nothing to do with "race."

Ethnicity is a valid way to describe social histories and social and symbolic identification, but it is not biology and most definitely is not race.

MOVING BEYOND THE MYTH

If, as a society, we can move beyond the myth of race as describing natural and biological units, then we can better address the inequalities that the race myth—and its concomitant, the social practices of racism—have created. The myth is strong, even in the face of resounding evidence against it. However, education and information (and access to them) are the main tools of myth busting. We will not move past this myth in this generation, or maybe not even in the next, but it is a possibility for the future of our society. As more and more of the myth-busting information discussed here becomes part of our social context, as children develop their schemata in the context of an accurate, information-rich social network, the effect on our cultural constructs and societal perceptions can be substantial. Some of these changes are already under way, but the forces maintaining the myth of race are many and massive, especially the current pattern of inertia, or maintenance of the status quo, in adults. We may find it very difficult to change our own views, or once changed, we may find it uncomfortable to speak up against this myth in many situations. Or, maybe we can try out the lyrics of the popular song "Your Racist Friend" by the group They Might Be Giants:[65]

> It was the loveliest party that I've ever attended
> If anything was broken I'm sure it could be mended
> My head can't tolerate this bobbing and pretending
> Listen to some bullet-head and the madness that he's saying
>
> This is where the party ends
> I'll just sit here wondering how you
> Can stand by your racist friend
> I know politics bore you
> But I feel like a hypocrite talking to you
> You and your racist friend

In order to move forward we all have to be active in the discussion about the reality of racism in the United States. We need to confront our racist friends, family, and society. This chapter contains the basic information and references leading you to more in-depth analyses of the myth of race and all the details that refute it. Many of our social norms and cultural constructs stand in our way; they support the inertia and patterns that maintain the myth or at least make it very difficult to challenge it publicly. However, once we have read this kind of information, we cannot be hypocrites, we must be myth busters.

5

Myths about Aggression

In his story of Dr. Jekyll and Mr. Hyde, Robert Louis Stevenson famously shows the dark side of humanity. The respectable and kind Dr. Jekyll devises a potion that enables him to bring to the surface his evil core. In Mr. Hyde, with his vile appearance and violent behavior, Jekyll sees that this alter ego "bore the stamp of lower elements in my soul." The concept that humanity has a violent and evil core is widespread; it is one of the oldest and most resilient myths about human nature. From historical and philosophical beliefs to current popular and scientific beliefs, the view that a savage and aggressive beast is a central part of our nature permeates public and academic perceptions. Given this view, it is a common assumption that if you strip away the veneer of civilization, the restraints of society and culture, you reveal the primeval state of humanity characterized by aggression and violence.

While there are many reasons for the resilience of this myth, the most powerful one is the simple fact that humans today can and do engage in extreme levels of violence and aggression. If you read the newspaper, visit online news sites, or turn on the television, you are guaranteed to come across some evidence of humans behaving violently toward other humans. While many animals aggressively hunt, capture, and eat prey, it is relatively rare for most animals to engage in intense, lethal aggression with members of their own species. Many social mammals display some intraspecific (within the same species) aggression and violence, sometimes resulting in death. A male lion might seriously injure another male lion in a fight over access to a pride of females, two rams might butt heads until one of them staggers away seriously hurt, or a male baboon might repeatedly attack a female in his group, wounding her and injuring her infant. However, these events, while aggressive and violent, are not the main ways in which the individuals in these species interact.[1] For the most part, death of opponents in these cases is neither the premeditated goal nor the outcome of the behavior. So, while intraspecific violence occurs, most species do not exhibit extreme aggression regularly and methodically. Humans are the only species that practice premeditated homicide and full-out war. That humans can, and do, participate in aggression and violence in ways that most other animals do not (and cannot) has led many to theorize that this aggression, this inner beast or demon, our Mr. Hyde, is part of human evolutionary heritage.

THE BEAST LIES WITHIN?

The selfish gene . . . accounts easily for selfishness, even killing.
And it has come to be applied with increasing confidence to human
behavior, although the debate is still hot and unsettled. In any case,
the general principle that behavior evolves to serve selfish ends
has been widely accepted; and the idea that humans might have
been favored by natural selection to hate and kill their enemies has
become entirely, if tragically, reasonable.

—Richard Wrangham and Dale Peterson (biological anthropologist; author)[2]

The myth of human aggression holds that we are indeed evolved to be killers, or at least aggressors who use the threat of violence as a major evolutionary tool. The mark of this evolved tendency toward aggression can therefore be found in our bodies and minds, especially those of men:

When we look at humans' bodies and brains, we find more direct signs of design for aggression. The larger size, strength, and upper-body mass of men is a zoological giveaway of an evolutionary history of violent male-male competition. . . . (Stephen Pinker, psychologist)[3]

There is the notion that aggression, the capacity for immense violence, evolved specifically because of the benefits it gave males, including an edge in competition with one another and between groups of males. Some make the argument for indicators of aggressive cores in our closest primate relatives and suggest that aggression and violence result in evolutionary benefits:

Thus, both the patterns of deadly violence in nature and the ethnographic record of simple hunter-gatherers clearly suggest that intra-specific human violence and the threat of it, while obviously undergoing transformations and varying in form through human history, are on the whole as old as humanity itself, indeed as old as nature. . . . (Azar Gat, political scientist)[4]

In short, the myth of human aggression is that humans (especially males) have a specific and distinct tendency toward aggression and violence and that this is patterned in our bodies and minds and arose due to evolutionary pressures of competition between men and between groups. If this were true, then aggression and violence must be a core part of who we are as humans because over evolutionary time those with the more aggressive behavioral patterns or traits must have defeated opponents and mated more successfully than those who were more pacific.

Testing Core Assumptions about Aggression

That humans (especially males) are naturally aggressive is a prominent myth about human nature. To bust this myth we have to test the core assumptions and refute them.

ASSUMPTION: *Human aggression (especially in males) is an evolutionary adaptation; we evolved to be aggressive, big-brained apes.*

TEST: Is aggression a trait that can be selected via evolutionary pressures? Is there evidence that humans, and our closest relatives, are evolved to be aggressive? That is, do aggressive behaviors and violence appear as central parts of our (and our closest relatives') evolutionary history, especially in males? If the answer to these questions is yes, the assumption is supported; if no then it is refuted.

ASSUMPTION: *The nature of human aggression is in our genes.*

TEST: Are there biologically identifiable and measurable factors that clearly demonstrate that aggression and violence are rooted in specific genetic or biological characteristics of humankind? If there are, then the assumption is supported; if not, then it is refuted.

ASSUMPTION: *Aggression results in "survival of the fittest." In an evolutionary sense being aggressive and violent, especially if you are a male, gets you more benefits.*

TEST: Do more aggressive, more violent, or more warlike males do better both in human society and in our closest relatives? If they do, then the assumption can be supported; if not, then the assumption can be refuted.

ASSUMPTION: *Humans, at their core, rely on aggression and violence more than cooperation and mutualistic interactions; everyone is aggressively just out for themselves.*

TEST: If studies show that humans are more successful when being selfish, aggressive, or violent in dealing with challenges and each other, then the assumption is supported. However, if cooperation or mutualistic interactions and nonselfish actions are more pervasive than aggression, then the assumption is refuted.

MYTH BUSTING: HUMAN AGGRESSION (ESPECIALLY IN MALES) IS NOT AN EVOLUTIONARY ADAPTATION

The myth about human aggression is relatively simple: the use of aggression is a core characteristic of human behavior and thus is a trait that has been shaped over evolutionary time. Looking at our closest relatives, the chimpanzees, and at humans across the globe reinforces this view. Richard Wrangham and Dale Peterson wrote a whole book laying out these themes: "We are apes of nature, cursed over six million years or more with a rare inheritance, a Dostoyevskyan demon. . . . The coincidence of demonic aggression in ourselves and our closest kin bespeaks its antiquity."[5]

The biologists Laurent Lehmann and Marcus Feldman draw a model specific to males (both in humans and other primates) to further elaborate on and connect the dots between aggression, maleness, warriors, and conflict between groups. Aggression leads to warriors, which lead to wars, which must be won to gain evolutionary benefits. Thus the best, most successful human males are aggressive by nature: "To this end, we assume that males, who are almost always the warriors in humans and higher primates, express two individually costly traits. The first trait causes an actor to be belligerent, which increases the probability that the actor's group goes to war and tries to conquer another group. The second trait causes an actor to be brave, which increases the probability that its group wins a war and conquers another group."[6] This is a convincing story. It resonates with what we hear on CNN and Fox news, read on the Internet, and possibly even experience at some point in our own lives; it seems like aggression is common and violence is pervasive in modern societies. There are wars in the Middle East and Africa, terrorism across the globe, and violent crime in the United States. Why would this be the case unless humans are by nature an aggressive species? It also seems that men are more aggressive than women. Most murderers are male, most rapists are male, and most people in prisons are male. Why would this be the case unless men were naturally more violent than women? It has to be true that at our core we indeed evolved into aggressive, big-brained apes (or at least men did). If it were not for laws, rules, and the constraints of society the law of the jungle—nature red in tooth and claw—would rule the day. Right? Wrong.

Aggression is not at the core of human nature, and human evolution did not produce a particularly violent species called *Homo sapiens*. The popular media and a portion of the academic literature may present this picture, but it is not an accurate representation of our evolutionary history or how we act in our everyday lives. Yes, we do hear a lot about murder and mayhem in the news on a daily basis. There are wars going on in multiple places on the planet and more men are in prison than women. However, is this what most humans do on a daily basis? Is there evidence that this is what most of us did in the past? Is aggression and violence more common and more important than other forms of relating to people, both now and in our past? If there were ten people murdered in New York City tomorrow, it would make the front-page headlines and be on news Web sites around the world. However, you are not going to see a headline the day after tomorrow asserting that "14 million people get along in New York today"—which is equally, if not more, important.[7] Along the same lines, it is true that there are a number of wars going on today, but are these wars best explained by a human tendency to be aggressive? What percentage of the seven billion people on this planet are actively engaged in war? Even in areas where we are at war, how many of the people in that area are acting aggressively at any given moment? Do we really see compelling evidence of a central role for aggression in the human past? Or are we overemphasizing and simplifying this reality?

Violence and aggression attract our attention more than nonaggressive or nonviolent interactions. They get more attention culturally and when they do occur they can have serious physical and health impacts. This can sometimes lead us to think that aggression and violence are more important (or at least impactful) than everything else we do, and thus must reflect a specific part of human nature. In this chapter we review substantial biological, psychological, primatological, and anthropological literature in an evolutionary context to demonstrate that there is no consistent support for the concept that aggression and associated violence are a specific human adaptation.

This does not mean that aggression occurs infrequently or that it is not important, but only that it is less central to humanity and our evolution as many think. In fact, most people do not realize that when we use the term "aggression" we are actually referring to many different things. It is not something that can be examined evolutionarily in the same ways as a hand or a tooth or hair color; aggression is much more complex.

Aggression is not a uniform or discrete trait

When talking about aggression we frequently treat it as a single thing. Think about what we mean when we say "humans are naturally aggressive" or "men are more aggressive than women." But is hitting someone the same as yelling at them or visually threatening them? Feeling really angry at someone or being very fearful of someone can produce aggressive reactions, but are these reactions the same thing? What about if you accidentally harm a bystander while directing aggression at another, are both victims receiving the same kind of aggression from you? In order to think about the "naturalness" of human aggression we first need to clarify just exactly what we mean by aggression, and whether or not we can think of aggression as a discrete trait.

In general, researchers who study behavior recognize that there is a wide spectrum covered by the term "aggression." One way to deal with this is using the dual concepts of *agonism* and *aggression*. Agonism is a broad category for behaviors associated with conflict and potential conflict, such as aggression, defense, and avoidance. Agonism is a general term for what we call antisocial behavior—actions by an individual or group that violate personal (or cultural) standards for appropriate behavior in a social context. Agonism may be tracked on a scale, from very mild antisocial behavior to more extreme antisocial behavior (this is the end that most of us think of as aggression). However, this terminology is not in wide use either by the public or by many researchers. So, for our purposes here, we can just think of aggression as a general term referring to a wide range of antisocial (agonistic) behavior.

It is important to note that the spectrum of aggressive behaviors is often thought of as a collection of responses that evolved in animals as part of a process to defend or obtain resources, as the psychologist John Archer explains: "The principle underlying an evolutionary functional analysis of aggression is a cost–benefit analysis. This has been applied to competitive aggression, for example in studies of the conditions under which animals show territorial aggression."[8] The basic assumption is behavior that helps protect an animal's life, defend against predators, or defend its own resources (food, space, young, etc.) would be favored by evolutionary processes so that over time most animals who displayed these behaviors (in the right contexts) would do better than those who did not. This means that aggression is a general category term for describing these responses and not a specific unit or trait. So the view is that particular behavioral patterns evolved in response to specific kinds of threats or conflicts that were common and regular in the lives of animals. This view largely assigns the display of aggressive behavior to the needs of survival and reproduction. Inherent in this notion of the origin of aggressive behaviors is the acknowledgment that there is an integral relationship between aggression and conflict and that aggression is generally defined by, and thought of as existing in, ecological or social relationships between organisms.

When we are specifically thinking about human aggression and aggressive behaviors there is a set of related definitions that are used. Generally, aggression is seen as being overt (measurable or obvious) behavior by one individual (the aggressor) that has the intention of inflicting physical or psychological harm on another individual (the victim). It is usually assumed that the intent to harm is immediate and not at some point in the future. Some researchers argue that the harm that arises from accidental or incidental behavior (either to bystanders or something that is not the aggressor's intended outcome) should not be considered as actual aggressive behavior. However, this basic definition does not really include threats and general hostility, which might also be relevant to our broader discussion or notion of aggression, especially considering the basic definition of doing either physical or psychological harm. When thinking about human aggression there is usually a separate definition for violence, which is seen as extreme, destructive, severe, and potentially cruel aggression. So violence is what resides at the most intense end of the aggression continuum. Many psychologists also make the differentiation between normal aggression and pathological aggression.[9] Pathological aggression is aggressive behavior that is particularly exaggerated, persistent, or expressed out of context.[10]

In studies of nonhuman animals, researchers often categorize distinct types of aggression to describe most effectively the specific causes and contexts of aggression and its constituent behaviors. The psychiatrists Allan Siegal and Jeff Victoroff assert that "it is empirically obvious and universally accepted that aggression is not a unitary phenomenon and that there is more than one type of aggression."[11] These types include fear-induced aggression (when an animal is prevented from escaping attacks or capture), maternal aggression (when mothers defend offspring), intermale aggression (a male in conflict with another male of the same species), irritable aggression (from exposure to a threatening or irritating item or event), sex-related aggression (where the sexual act includes aggressive actions), predatory aggression (when an attack response is triggered by the presence of a prey item), and territorial aggression (when an intruder enters into another animal's defended territory). However, most researchers also agree that these types are not exclusive and that any single aggressive interaction can potentially have multiple types of aggression in it. The aggressive behaviors can fall at different places along the continuum of mild to severe. Many researchers also note a specific difference between the behaviors involved in predation (hunting and eating another species) and the types of aggression that occur between members of the same species.[12]

In practice there are some general correlations between these categories and human aggression. However, human aggression can be hard to classify specifically into these types as it often involves much more complex planning, more varied social and cognitive contexts, and occurs in both one-on-one conflicts and in larger-scale interactions.[13]

So, human aggression consists of a suite of behaviors, with evolutionary origins in defense and resource acquisition, which reside on a continuum of severity. We can divide up these behaviors into different types of aggression, with different intents and contexts. Basic categories are proactive or predatory aggression and affective or responsive aggression.[14] Responsive aggression covers all aggressive behavior that is reactive, defensive, and generally hostile but is associated with fear or response to threat. This is what emerges in defensive interactions, in protecting your children or loved ones, or when you are surprised or shocked and react aggressively. Proactive aggression is premeditated and potentially coordinated. This is when you set out to harm someone, focus in on a specific kind of harm, or plan a series of harmful events. One can see how these two types of aggression might use similar behaviors but the reasons they occur and the outcomes from them can be really different.

In general human aggression is largely context-dependent, and non-pathological aggression is frequently associated with states of anger or fear. The psychologist and aggression expert John Archer reminds us that the human emotion of anger comes from an evolutionary ancient interaction of the brain, hormones, and behavior (often called the flight or fight response system), thus linking human aggression in some ways to the more generalized animal aggression. While anger is typically associated with aggression, similar circumstances can elicit both anger and fear, which can result in different types of aggression (reactive and proactive). In animals there is usually a clear separation in the brain-behavior pathway between anger-induced and fear-induced aggression, with actual physical aggression being more restrained in anger-based behavior than in fear-based behavior. However, in humans (relative to other animals) there appears to be a wider range of ways in which aggression can be expressed and the distinction between anger-induced and fear-induced aggression is not always so clear cut.

Long-term studies (extending over two or three generations) show that aggressive behavior patterns are moderately consistent in individuals over their lifetimes, at least for physical aggression. That is, just over a third of children in two studies (in the United States and Finland) who demonstrated high aggressive behavior levels at around eight years of age also displayed these levels as adults. Interestingly, the same is true for those children who displayed very low levels of aggression. Not surprisingly another strong correlation with childhood and adult aggression was the aggressiveness of the adults parenting the children. This pattern, where context, learning, and childhood experience and environments are related to the adult expression of aggression, is characteristic of many cultures around the planet.[15] Interestingly, separate but related studies with twins also suggest that there may be some effects of biological similarities, with identical twins being slightly more likely to share these patterns than nontwins, even when raised separately.[16] This suggests that the expression of aggressive behaviors in humans involves integration of biological and cultural systems.

The psychologists C. Nathan DeWall and Craig A. Anderson recently summarized the basic premise of a general aggression theory.

They state that the way aggression is displayed depends primarily on the cognitive and experiential factors of the person displaying it. They suggest that the way in which humans perceive and interpret the environment and other persons in it, their assumptions and expectations of their situation, how they believe others will respond to events, and how much they believe in their own abilities to respond to an event, all affect how aggression will be displayed in any given situation. Basically the concept is that how people "see" and experience the world (remember the discussion of schemata and culture in chapter 2) shapes how they perceive events, objects, people, and social interactions as relevant or irrelevant to aggression.[17] So an act of aggression itself results from the basic capacity for exhibiting these types of behaviors (which are physiological responses related to our evolutionary history and biological present) combined with the convergence of an individual's experience, history, perceptions, and a given particular situation.

Given what we have reviewed here, we can safely say that aggression, as humans exhibit it, is not a single, unitary trait, or even an easily described physical or behavioral system, that can be shaped by evolutionary pressures in the same way as a trait such as a tooth or a femur (leg bone). However, what about specific patterns that result in particular types of aggression? If there are clear and robust patterns where we can see humans use aggression in the same ways and contexts over evolutionary time or across cultures, then we could say that there has been some form of evolution of aggression, maybe especially in males, over time. So do particular types or patterns of aggression appear regularly as a core part of being human due to human evolutionary histories and do these appear more commonly in males? To answer this we need to ask about our closest relatives, our evolutionary record, our histories of aggression, and our modern behavior, especially differences between men and women.

What can other primates tell us about human aggression?

The basic behaviors used in aggression have their origin in the flight or fight response. This is a physiological system common in all animals, and developed in a specific way in mammals. When something potentially dangerous happens to a mammal its body responds very quickly with certain hormonal, heart rate, and behavioral actions. For example, say you are a zebra eating grass on a savannah and you look up to see a lion running at you. Your eyes and brain immediately send a series of signals to your heart, muscles, and adrenal glands whose actions then allow you to turn and run, extremely quickly and effectively, without thinking about it.[18] This same scenario, in a slightly different situation, can also result in fighting. Say a zebra is cornered by a pack of wild dogs against a cliff face. The same system of brain, heart, and adrenal glands will act, but in this case the zebra will use the energy and muscle power to fight the dogs instead of fleeing. The bottom line is that in animals we know where responsive aggressive behavior comes from.

It turns out that the bodily systems involved in proactive aggression are basically the same as in responsive aggression; however, they are stimulated by, and used, in different situations. In complex social mammals (primates, wolves, whales, dolphins, etc.) proactive aggression can be especially important in social contexts. In fact, it is often argued that by looking at humans' closest relatives, the primates, we can see some of the evolutionary origins for human proactive aggression and violence. Is this the case?

Anyone who watches nature documentaries or programs on primates would probably agree with political scientist Azar Gat's assessment: "Wide-scale intra-specific deadly violence has been found to be the norm in nature, including among our closest cousins, the chimpanzees."[19] Nature documentaries often show monkeys and apes spending a great deal of time fighting, in addition to having sex and grooming one another. However, Azar Gat is wrong. While fighting, having sex, and grooming are very important to primate societies, such social interactions, especially aggressive ones, are not at all common. Most primates actually spend the vast majority of their time looking for food, eating, and resting. Even the most socially active of primates (macaque and baboon monkeys and chimpanzees) rarely spend more than 20 percent of their time in social interactions. Across the primate order the average total time spent in social interactions ranges from 5 to 10 percent of total active time. Humans are by far the most socially active primates. However, this does not mean that social interactions are not important; they are the glue that hold primate societies together and make up some of the most interesting and central aspects of primate lives. If we look at how social interactions break down we see something very interesting: primates spend most of their time in prosocial behavior; aggression, especially severe aggression, is very rare. In direct opposition to Gat's assertion, a broad overview of the primatological literature, including hundreds of species and tens of thousands of observation hours, demonstrates that intraspecies violence resulting in death is extremely rare, and decidedly not wide scale, in primates.[20]

This does not mean that primates do not have conflicts and fights; in fact low-intensity aggression (mild agonism) is a part of daily life. If you spend a day with a group of primates you are likely to see a few small fights, an array of mild threats, a few slapped hands, and even a bite or two. But actual wounding and other types of severe aggression are rare in the primate world. The anthropologists Robert Sussman and Paul Garber examined published reports for over sixty species of primates and found that the overwhelming majority of interactions were prosocial and cooperative. Agonism (the entire spectrum of aggressive and conflictual behaviors) made up less than 1 percent of all social interactions, with instances of observable aggression ranging from just under one an hour to less than one per hundred hours across different species.[21]

This does not mean that aggression is not important. It is. Positive social relationships are so central to primate societies that the possibility of aggression, even rare aggression, can elicit a whole suite of behavioral patterns that have evolved to ameliorate, or fix, the damage done by aggression between members of a group. This pattern is called the valuable relationships hypothesis and is strongly supported by a range of research projects.[22] So while aggression is quite rare, it can be important. If this is the case, then we would expect to see that even when aggression is displayed, on average there will be mechanisms at play to minimize the social impact of such aggression. As psychologists Paul Honess and C.M. Marin note (in a review of primate aggression), "it is now clear that due to the high potential cost of physically aggressive behavior, most animal aggression occurs in often highly ritualized contexts designed for maximum effect and minimum risk."[23] So, aggression occurs and is important, but as a general rule frequent, damaging, and violent aggression is not a pattern seen in the animal kingdom or among our closest primate relatives.

However, there are two areas where we can look for specific patterns of aggression in other primates that might have something to tell us about human aggression: male aggression toward females and violent confrontations between communities of chimpanzees.

If you look at frequency of aggression in primates by sex (male or female) it tends not to vary that much. However, if you break down this already relatively infrequent behavior by intensity, you find that in a majority of species males engage in more serious aggression than females and tend to have higher rates of wounding.[24] You also find that females, in some species, engage in aggression slightly more frequently (but at a lower level of physical harm) than males. This pattern is probably best attributable to two causes: females are more social than males and males are larger than females. The former simply reflects the fact that in many, but not all, primate species the majority of social interactions go on between females and females and their young. Adult males generally have lower rates of social interactions overall. Because aggression is relatively rare but does require social partners, we find that those with the highest degree of social interactions tend to have higher rates of aggressive behavior (thus females can be expected to be slightly more frequently aggressive than males). But when males are aggressive, the impacts can be more serious.

In most primate species males are larger than females. Males often have larger canine teeth, and combined with their body size, have a greater potential to harm females than females do to harm males (in one-on-one aggressive conflict). This means that the outcome of males engaging in severe aggression can be more harmful (physically) than for females. There are some researchers who argue that this pattern is reflective of an evolutionary adaption for males to use aggression to control females. While it is true that in some species (like chimpanzees, orangutans, and some baboons) males do seem to use aggression to coerce females to stay near them, or even to mate with them, there are also many other species (gibbons, many macaques, marmosets and tamarins, and Asian leaf monkeys) where males are not able to use aggression to coerce females at all. In fact in many species females can group together to form coalitions with which to resist male attempts at coercion or aggression.[25] This suggests that there is a wide range of male use of aggression as a social tool in the primate order, and that there is not a specific pattern that characterizes all primates that we could point to as a shared evolutionary basis for patterns of aggression in humans.

What about chimpanzees?

Clearly a primate-wide survey does not give any strong insight into human aggression beyond what we already know about animals' basic patterns and the relationship between increased social interactions and the increased potential for aggressive interactions. But what about our closest relatives, the chimpanzees? Can a closer look at them tell us more about our evolutionary past? Possibly. If there were specific and marked patterns of aggression that we see in chimpanzees and in humans then we might be able to propose that these similar behaviors reflect a shared adaptation which characterized our last common ancestor. After over fifty years of focused study on a number of chimpanzee communities there is no doubt about one thing: chimps can be really aggressive, as several primatologists attest. David Watts reports that "chimpanzees stand out quantitatively and qualitatively among the nonhuman primate species in which lethal coalitionary aggression is known."[26] Rebecca Stumpf observes that "chimpanzees are very territorial, and intercommunity interactions are usually hostile and sometimes deadly."[27] But Christophe Boesch cautions us not to overemphasize the level of aggression in chimpanzees:

> Lethal intergroup aggression in chimpanzees has been proposed to present similarities with primitive warfare in human populations, based mainly on observations that both species regularly use large male coalitions, systematic patrolling of territory boundaries, and violent killing of adults from neighboring groups that can lead to the annihilation of a whole group. . . . However, the sometimes dramatic nature of intergroup aggression in chimpanzees has led to under appreciation of the fact that such lethal violence represents only the minority of the encounters between communities in wild chimpanzees.[28]

There are two species of chimpanzee, the chimpanzee and the bonobo (same genus, *Pan*, but different species, *troglodytes* and *paniscus*, respectively). The chimpanzee, *Pan troglodytes*, has four types (or subspecies) that can be roughly divided into western and eastern African forms, with *Pan paniscus* found in between them. Chimpanzees are big brained and very complex social primates. They make and use tools, they have localized social behavioral patterns that some refer to as a culture, and in captivity they can be trained to communicate with humans using rudimentary forms of symbolic representation (types of sign languages).

Here is what we know about chimpanzees and aggression:

1. All chimpanzees live in large communities with many males, females, and young. Most of the time the communities are broken up into subgroups that range across the overall territory of the community. For eastern chimpanzees the whole community is not often together in the same place at the same time, while

for western chimpanzees and bonobos large groups and whole-community gatherings are more common.

2. Both eastern and western chimpanzee males in a community occasionally get together to patrol the boundaries of their territory (sometime females join in, but very rarely among eastern chimpanzees). When these males encounter individuals from neighboring communities (between 3 percent and 12 percent of the time) they often act violently and the interaction can result in death. These aggressive attacks seem to be based on an imbalance-of-power scenario, with the patrolling group attacking only if they greatly outnumber the individuals encountered.

3. Chimpanzee males and females (primarily eastern chimpanzees) have been observed killing infants from their own community as well as from other communities.

4. In eastern chimpanzees males are dominant over, and particularly aggressive toward, females and often are able to coerce them using aggression and threats. In western chimpanzees, while males are more or less dominant they are less likely to use aggression to attempt to coerce females (and when they do they are less successful).

5. Bonobos do not engage in border patrols, lethal intercommunity violence, or infanticide, males do not coerce females, and females are dominant to males in many interactions.[29]

What does this tell us about human aggression? Unfortunately, not much.[30] Humans are equally related to the two species of chimpanzees and those two species are significantly different when it comes to the types and patterns of aggression they display. On the one hand, *Pan troglodytes* males can be really aggressive to females but *Pan paniscus* males rarely start fights with females (and when they do, they often lose). The eastern and western chimpanzees' border patrols and intercommunity lethal violence might be similar in some ways to human behavior. Human males can, and do, form groups and attack other groups of humans in many different circumstances, occasionally involving pretty severe or even lethal aggression. But is human feuding, homicide, and war really comparable to the eastern and western chimpanzees' behaviors? Some researchers have argued that this is in fact the evolutionary origins of war (which we examine in the following section).[31] However, most anthropologists and biologists will not take the comparison to that conclusion. Christophe Boesch notes that "intergroup violence is general in chimpanzees, but aggression intensity and the role of females vary considerably among populations. Thus, multiple factors may account for the evolution of intergroup violence in this species, and we need more demographic data on neighboring communities if we want to understand intergroup conflict dynamics."[32]

And then of course there are the bonobos, who do not display this kind of severe aggression. Even ignoring the bonobos, the variation across the eastern and western chimpanzee subspecies makes it a bit difficult to substantiate direct assertions about shared patterns between humans and chimpanzees, especially about the roles of aggression and male-female interactions. This is compounded by the fact that, while the lethal attacks between groups are important, they are really very rare and make up a very small part of the overall lives of all chimpanzees, and many chimpanzees may go their whole lives without encountering such a situation.

This overview of chimpanzee data suggests that we probably share with them the potential for severe aggression between groups and male coercion of females. But even chimpanzees themselves vary in how this plays out across species, subspecies, and communities. There is no smoking gun reflecting an evolved system of lethal aggression and violence in the shared chimpanzee and human history. There is, however, an interesting association between being really socially complex, living in large and dynamic communities, and having varied and complicated social lives, and the potential for variation and complexity in aggression. Maybe what we can take from comparisons to the primates, and especially to chimpanzees, is that social complexity, history, and context are important in understanding why a certain form or pattern of aggression occurs in a group. Humans are not chimpanzees, but as the anthropologist Jon Marks puts it, we are "biocultural ex-apes." That is, we are a branch of the evolutionary lineage that includes apes, such that there are going to be aspects in our bodies and behaviors that might reflect that shared heritage, but at the same time our adaptive zone—culture, and its complexity—makes us very different, in both the potential and the actual behavior that we express, from our chimpanzee cousins.[33]

War is not part of our evolutionary heritage

What about warfare and violence in the human evolutionary past? Today, at any given moment there are multiple wars ongoing at various locations on the planet; in all likelihood someone was just killed in a war-related act of violence as you read this. However, in the same moment, billions of people were not exposed to, or involved in, aggression at all. So, while peaceful interaction is a lot more common than war, and more widespread at any given moment, war results in a more immediate and dramatic outcome than peace—death. Warfare is a part of modern humanity and it impacts lives, so the evolutionary history of warfare is important and might be able to give us insight into human aggression.

This leads us to the question: are war and aggression connected? One might assume the answer is yes because war involves killing and we assume that aggression is largely necessary to kill.[34] There are many who assume that it is indeed the human "killer instinct" that enables us to even conduct war. If war is an ancient pattern in human evolution we expect that certain kinds of aggressive behaviors would have evolved, and be present, in modern humans in response to this long-term need for the ability to engage in complex lethal aggression. So the real question is this: Is war a part of our evolutionary history?

While there are countless definitions of war out there, we will stick with a relatively simple one: war is organized lethal violence by members of one group against members of another group. This definition is important because aside from possibly chimpanzees (especially eastern ones) the specific behavior of war is unique to human beings. Groups of monkeys occasionally fight over fruiting trees; ant colonies will run into one another and fight on occasion, and many other types of animal groups engage in conflicts, but none of them are planned, organized, and lethal with regularity. This definition also differentiates war from homicide (single events where an individual is killed by another).[35] The distinction is important when we are thinking about aggression, as killing another can result from responsive (defensive) aggression or from proactive aggression in a wide array of contexts (fear, threat, surprise, anger, etc.). However, homicide is a one-off event without the kinds of important intragroup coordination, planning, and implementation required for warfare and does not require the evolution of a system of aggressive behaviors.

As anthropologist Brian Ferguson contends, "If humans had an inborn predisposition for violent conflict, then they should have been war makers since, or even before, they became human."[36] Yet if you review the published information on the fossil record of humans and potential human ancestors from about six million years ago through about 12,000 years ago, you are provided with, at best, only a few examples of possible death due to the hand of another individual of the same species.[37] Most of the early cases can be attributed to cannibalism (though it is not clear if this resulted from within-group or between-group actions) or generalized injuries that may have come from fights. The earliest potential cases of injuries resulting from acts of war are a few instances of possible defleshing of body parts after death in *Homo erectus* and Neanderthals. Clear evidence of actual deaths from spears and bows and arrows is only found in modern humans about 8,000 to 12,000 years ago. From that point on, most archeological records show some examples of individuals having died at the hands of others.[38] So despite misleading assertions such as Gat's that "comprehensive examinations of large specimens of fossilized human bones have concluded that at least some of them were injured by human violence," examination of the human fossil record supports the hypothesis that while some violence between individuals undoubtedly happened in the past, warfare is a relatively modern human behavior (12,000 to 10,000 years old).[39] This suggests that the fossil evidence supports the assertion that war is not part of our deep evolutionary past.

In his 2008 article entitled "10 Points on War" Ferguson lays out ten key factors, based on his three-decade study of human war.[40] These are the minimal set of points we need to consider when engaging in a discussion about the role of war as it relates to understanding human aggression:

1. Our species is not biologically destined for war.
2. War is not an inescapable part of social existence.
3. Understanding war involves a nested hierarchy of constraints.
4. War expresses both pan-human practicalities and culturally specific values.
5. War shapes society to its own ends.
6. War exists in multiple contexts.
7. Opponents are constructed in conflict.
8. War is a continuation of domestic politics by other means.
9. Leaders favor war because war favors leaders.
10. Peace is more than the absence of war.

The zoologist Robert Hinde reinforces the complexity of warfare when he states that "in major international wars people do what they do mainly because it is their duty in the role they occupy; combatants in institutionalized wars do not fight primarily because they are aggressive."[41] Both Ferguson's and Hinde's statements demonstrate that war is neither an easily defined phenomena nor something to be simply assessed as an indicator of an evolved human tendency for aggression. In fact to try and disentangle those things that go on in war from their social, historical, political, and situational context is probably a pointless task. As the late biological anthropologist Philip Walker, a prominent expert on indicators of violence in the fossil record, reminds us: "It is useless to try and pin down a single natural or cultural cause for violence in the past, and it is that very complexity in the causes of violence that should be our final lesson."[42] War is not a part of our evolutionary history, but it is obviously a major part of our current potential.

Males and females differ in aggression, but not the way you think

In nonhuman primates we do see some sex-based differences in aggression. Males are generally larger, with larger canine teeth. Therefore, males potentially may demonstrate more damaging physical aggressive behavior. Overall males and females both participate in aggression, but the rare instances of severe, or even lethal, aggression are primarily displayed by males (with targets being both males and females). Low-grade aggression, or mild agonism, is slightly more common in females than males. It has been argued that this pattern of size difference in many primates is the result of sexual selection. That is, over evolutionary time there has been a favoring of larger males because males engage in competition with one another over females.[43]

Like other primates, humans do exhibit a fairly consistent suite of physical and physiological differences between the sexes. In any given population men tend to be between 7 percent and 15 percent larger than women.[44] Males, on average, have a higher muscle density per unit area and more upper body strength than females (even at the same body size), and while the cycling hormones in males and females are the same, there are differences between the sexes in the levels and patterns of some of those hormones (such as prolactin and testosterone).[45]

If you look broadly across the published research on human aggression there are some patterns of differences between males and females. Unlike when talking about primates or other animals, in humans when we talk about male and female behavior we are primarily talking about gender. For anthropologists, the major distinction between sex and gender is one of societal context. Sex refers to the biological differences between men and women and gender refers to the overall differences in perception and actions. Just as we are enculturated into the society in which we grow up, we are also engendered (literally "to become a gender"). This topic is expanded on in chapter 6, but suffice to say that in this discussion when we talk about sex differences we are referring to differences in biology and when we talk about gender differences we are referring to behavioral patterns and perceptions.

While nearly all studies demonstrate much overlap between the sexes in all types of aggression, there are patterns of difference that seem to be relatively consistent. These patterns are what we can call the basic sex differences in aggression between males and females. Remember, however, that these are average patterns and that any given individual might deviate from these expected trends due to a wide array of situations. For aggressive encounters between same-sex individuals or between a known sex and an unspecified sex as the target, the following patterns emerge.[46]

1. Males generally display more physical aggression than females.

2. Males display slightly more verbal aggression than females.

3. There is no difference in the rates of anger displayed by males and females.

4. As young girls, females display more indirect aggression than males, but this drops to equal levels by adulthood.

While all groups of individuals show a range of variation in the display of aggressive behavior across most studies, there is a greater variability in males' display of physical aggression than in females' physical aggression (but this is not true for other types of aggression). These differences appear to begin early in life, starting around two years of age. Rates of physical aggression peak right about this same time (two to three years of age) and then decline until about nine to eleven years of age in both males and females. At two years of age and even after the decline, overall rates of physical aggression are generally higher in males. This pattern has led many researchers to conclude that the expression of aggression in both males and females is heavily influenced by social contexts as well as their interactions with both sexes' growing bodies and physiologies. The bottom line is that, on average, males do seem to show more physical aggression than females in same-sex conflicts and in general use of aggression.

Interestingly, studies that actually follow individuals from childhood to adulthood (and beyond, sometimes following two or three generations) show us that it is not always possible to predict adult aggression by looking at children; however, there are some sex differences here too. At least three long-term studies of children into adulthood reveal that nearly half of males who displayed high rates of aggression in early childhood retained that aggression pattern into middle adulthood, whereas only 18 percent of females who displayed high aggression in early childhood maintained it into middle adulthood. However, both males and females who displayed low aggression during early childhood had the same pattern of retention into middle adulthood (38 percent and 36 percent, respectively). This result suggests that there might be differential socialization for males and females (allowing more high aggression in males and dampening the expression of aggression in females), resulting in the different trajectories of retention of patterns of high aggression.[47] This is especially interesting given that there does not seem to be a link between physical aggression and the testosterone spike in males at puberty, which might have been an explanation for the maintenance of increased high aggression into adulthood by males.[48]

One area of aggression that seems particularly relevant to our overall questions is that of aggression between male and female partners. There are a number of studies that examine these interactions specifically, looking for any aggression differences between males and females in partner relationships. The results are a bit surprising. Unlike the general patterns of aggression noted above, there was little to no difference between males and females in the use of physical aggression in opposite-sex couples. In fact, overall use of physical aggression in this context is slightly higher in females than males. However, if you only look at samples involving extreme domestic violence (physical aggression resulting in injury) then males are more likely to do it. If you just look at the self-report data (fifty-eight different studies in this case) they show that women were significantly more likely to commit most acts of physical aggression, except for using weapons and actually beating up partners (severely, beyond the use of slaps, pushes, throwing things, and some hitting). Overall in these opposite-sex interactions, women were more likely to receive serious injuries than men. However, this data set is largely from Western societies, and John Archer notes that this relative equality between the sexes in acts of physical aggression is confined to nations where women have higher levels of societal power and the magnitude and direction of the sex difference followed a measure of societal gender empowerment and beliefs about gender roles. In a meta-analysis of cross-cultural partner physical aggression Archer demonstrates a direct tie to the gender empowerment index: the lower the female empowerment rating the higher the male bias in partner-based physical aggression.[49]

Finally, overviews of crime statistics from North America and Europe indicate that males are more likely to be involved in violent crimes and same-sex homicides. Nearly 80 percent of those involved in weapon-based crimes are male and about 97 percent of same-sex homicides involve males. There is a great deal of social complexity in explaining these figures, but in general, the most robust explanations mix social, economic, historical, and experiential factors in with the possibility that there may be certain evolutionarily influenced tendencies toward risk taking in young males.

What can we summarize from these data? On average, males appear to engage in more aggression than females, especially physical aggression when in conflicts with other males. But in aggression between opposite-sex partners there is very little difference in the use of physical aggression, with females being slightly more likely to use it. However, in this same context physical aggression by males can cause more damage, and in scenarios where real physical injury has occurred, males are more likely to have been the culprits. This basic outcome should not be that surprising given the likelihood that the males in any interaction are going to be 7 to 15 percent larger than the females, have a higher muscle density and greater upper body strength, and are more likely to have kept up high-aggression behavior if they exhibited it in childhood.

What does this tell us about males evolving to be more aggressive than females? It is not clear. There is not a simple picture of aggression and violence in men and women. Males do appear to be more aggressive in some contexts and females in others. The outcomes from male physical aggression and males' participation in antisocial aggression (crime and such) are more serious than those of females. However, regardless of sex, the majority of humans do not engage in violent aggression with any regularity (or even at all). But the confluence of physical size and strength with societal patterns and contexts and the experiential histories of individuals make it very difficult to discern whether there is an evolutionary pattern or not.[50] Humans of both sexes use aggression and males, being larger, can have more impact in physical aggression and are also more likely to be involved in violent crimes. Does this imply that males have evolved to be more aggressive than females? No, there is too much overlap and complex social contexts (childhood experience, gender empowerment index, social structure and crime, availability of weapons, etc.) to be able to make a clear line from what males can do in the use of aggression to a specific evolutionary benefit. However, we can point out that there are important differences in the outcomes of male versus female aggression due to aspects of our evolutionary histories: the male's larger body size and the higher likelihood of young adult males participating in risk-taking behavior.

HUMAN AGGRESSION, ESPECIALLY IN MALES, IS NOT AN EVOLUTIONARY ADAPTATION

Aggression, as humans exhibit it, is not a single trait, or even an easily described physical or behavioral system. It is not a thing that has evolved as a package, but rather as a suite of behaviors with a dynamic and complicated range of expression. There is a wide range of aggression in primates, but no specific pattern that characterizes all primates (or male primates) or acts as a shared evolutionary basis for patterns of aggression in humans. Rather, looking at the primates tells us that social complexity, history, and context are really important in understanding why a certain form or pattern of aggression occurs. Chimpanzees, at least some of them, do exhibit some very aggressive behaviors, but others do not. Which are the best models for human evolution? Neither. Humans share a branch of the evolutionary tree with chimpanzees, so aspects of our bodies and behaviors might reflect some of that shared heritage, but our separate evolutionary histories, culture, and overall complexity make us very different than our close chimpanzee cousins. If anything, the variation in chimpanzees is an important take-home message for understanding the range of potential variation in human ancestors.

Aggressive behaviors and violence do not appear as central parts of our evolutionary history. The fossil record demonstrates no evidence of large-scale aggression (war) or of specific and widespread patterns of interindividual aggression prior to about 10,000 to 12,000 years ago. But over the last 10,000 years or so we have increasingly participated in organized lethal violence, pitting members of one group against members of another group. So while war is not a part of our evolutionary history, it is certainly part of our current potential.

Finally, aggressive behaviors and violence are not central parts of male as opposed to female behavior. But the overall picture regarding aggressive behavior and the sexes is not particularly clear. There is no behavioral evidence that being aggressive is evolutionarily more likely for males than females, and while males appear to be more aggressive in some contexts, females are in others. Males are more likely to be aggressive than females in same-sex aggression, and there are differences in the outcomes of male versus female aggression due to male body size.

Overall, looking at the available evidence from the fossil record, our primate cousins, and current data on human aggressive behavior, the myth that there is an evolved human tendency toward aggression is busted.

MYTH BUSTING: THE EXPLANATION AND NATURE OF HUMAN AGGRESSION ARE NOT FOUND IN OUR GENES

Studies of nonhuman animals have shed light on the behavioral, neurobiological and molecular mechanism of aggression. Relating this mechanism to the human condition is difficult because aggressive behaviors are diverse.

—Randy Nelson and Brian Trainor (neuropsychologists)[51]

. . . no work has demonstrated that non-pathological humans have an inborn propensity to violence, and comparisons of males and females are uniformly complicated, qualified, and debatable. The growing appreciation that genetic expression occurs within a system of biological systems, all with environmental inputs, greatly complicates key issues. We are far from being able to clarify how and the extent to which inborn biological variables affect human or male aggressive behavior.

—Brian Ferguson (anthropologist)[52]

Now researchers have found signs . . . in the genes of our primate cousins . . . a team of geneticists traced one genetic variant to an allele that predisposes men to aggressive, impulsive, and even violent behavior.

—Ann Gibbons (journalist)[53]

One of these quotes is not like the others. From the preceding section it is readily apparent that human aggression is a very complicated theme, and yet the popular press and much of the public (and some academics) hold the belief that there is a specific biology or a genetic basis for aggression, especially in males. Identifying the genetic key to aggression is not possible, because it does not exist. Identifying the genetic and broader biological infrastructure for aggressive behavior is possible. This is a key distinction that is important to clarify before we talk about the physiologies and genetics associated with extreme or atypical aggression in humans.

Where Normal Aggression Comes From

We have reviewed the basics of the flight or fight response, but in actuality we know a great deal more about how the body's systems are involved in aggression. When we talk about aggression we are talking about actual behavior and, in general, behavior is initiated and controlled by the nervous and endocrine systems (the brain and hormones). Most of what we know about the brain's involvement with aggression comes from studies when things go wrong, but we do have some basic knowledge about the expression of aggression in nonpathological situations.

It is pretty clear that in humans two parts of the brain, called the prefrontal cortex and the dorsal anterior cingulated cortex, are centrally involved with the expression of behavior, especially aggression. The prefrontal cortex is linked to other behaviorally important brain structures called the amygdala and the hypothalamus. In general, these parts of the brain receive a variety of inputs from other areas of the brain (vision, smell, touch, pain, sound, memory, language, etc.) and then interact in a sort of feedback loop to stimulate other bodily systems (like hormones, neurotransmitters, and muscles) into action. The prefrontal cortex does a bunch of other things as well, including playing central roles in introspection, recognition of emotions, regulation of emotions, and detection of conflict situations and acts as a trigger to initiate a variety of other neurological systems in regard to social interactions. The dorsal anterior cingulated cortex seems to be involved in the regulation of responses to anger, pain, and social rejection. From brain imaging studies we know that individuals who are particularly aggressive often show lowered neuronal activity, reduced glucose metabolism, or even reduced density of gray matter in the prefrontal cortex than those who are not as aggressive. Studies of individuals who have received brain damage to the prefrontal cortex and amygdala reveal that they demonstrate more impulsive and antisocial aggressive behavior or have lowered abilities to control the expression of aggression.

Additionally, shock therapy in the 1950s and 1960s directed at the amygdala and prefrontal cortex resulted in lowered overall arousal rates and severely reduced aggression. In short, there are multiple studies which all point to the action of the prefrontal cortex, the dorsal anterior cingulated cortex, and at least the amygdala, as important areas for understanding the biological infrastructure of aggression.[54]

There are a suite of molecules (called neurotransmitters) produced by the body which directly interact with these regions of the brain and are involved in the expression of aggressive behavior (among other things). They are the 5-hydroxytryptamine receptors (5-HT for short and involved with the neurotransmitter serotonin), the neurotransmitter and neurohormone dopamine, the metabolic enzyme monoamine oxidase A (MAOA), and a variety of steroid hormones such as testosterone, other androgens, and estrogen. None of these are a smoking gun for the origin and expression of aggression in general, but some of these are implicated, to some extent, as playing a role in the emergence of particular types or patterns of aggression.

None of the information available on the neurological infrastructures for aggressive behavior implicates these systems as uniquely responsible for aggression, as specifically selected over evolutionary time to serve a function in aggression, or as generating a fixed outcome in the expression of any aggressive behavior. All of these systems have many other functions, are only implicated as components in the expression of aggression, and are wholly tied to, and shaped by, the lived history, social context, health, and daily lives of the individuals whose bodies they are a part of. There is no gene or system in the body that can be labeled "for aggression."

My genes did not make me do it, but they helped

Genes don't cause aggression in humans but if they are involved in the production of molecules that interact with the body's systems that are part of the expression of aggression, we could consider that genes can at least play a part in aggression.

What exactly do genes do? We know from chapter 3 that genes are basically just stretches of DNA that contain the code for either the production of a protein molecule (or parts of a molecule) or the regulation of other genes or of themselves. Genes come in multiple forms (alleles). While genes contain codes for proteins and their regulation, the relationship between genes and complex molecules like neurotransmitters and hormones is very complicated. The possible relationships between genes and behavior are even more so (see chapter 3 on gene-trait relationships). Let's take a look at what we know about some of the better-studied molecules implicated in aggression and see if we can establish more clearly that aggression is not "in the genes" but rather that genes, neurotransmitters, and hormones play a part in affecting the ways in which aggression is expressed.

Monoamine Oxidase A (MAOA)

Dubbed the "warrior gene" in the press, the gene that codes for monoamine oxidase A (MAOA) has recently been a central player in the study of genetic influences on aggressive behavior.[55] This gene is found on the X chromosome (the one you get from your mother), so that males have only one copy of it and females have two (males are XY and females are XX). The gene product, the enzyme monoamine oxidase A, interacts with the neurotransmitters serotonin, dopamine, and norephedrine, regulating their release and breakdown so that once they do what they are supposed to do they don't build up or interact with other receptors, causing problems for communication between parts of the brain.

It turns out that there are at least four different common alleles for this gene that have the effect of increasing or decreasing the amount of MAOA produced. Lowered amounts of MAOA in the brain in some mice, rhesus monkeys, and humans, under certain conditions, is associated with increased aggression and reduced ability to control impulsive behavior.[56] A noticeable number of mice, monkeys, and human males who had the low MAOA production alleles and who experienced severe social and/or physical trauma or abuse during early childhood development were more likely to express heightened aggressive and antisocial behavior as adults. Some low-MAOA humans also score higher on self reports of aggressive and violent behavior. However, in some cases the high-producing alleles are correlated with aggressive behavior in male children. In a famous case of a Dutch family who have a very rare allele where no MAOA is produced at all, three males exhibited extreme aggressive and antisocial behavior. Now, not all individuals with the low-expression alleles exhibit this kind of aggression, not even all of those with the low-expression alleles who had traumatic or abusive childhoods. In addition some of the high-expression allele carriers exhibited high aggression.

All of these studies were conducted on males because it is much easier to discover which alleles are acting, as males only have one copy of the gene. In females it is more difficult to identify the actual action of the gene because they may have two copies but only one is actually active; determining which one that is can be very difficult. Thus, while this gene is often invoked as an example of a male biological basis of aggression, there have been no in-depth studies of this gene in females so we do not know if it functions the same way. We should note that the enzyme MAOA operates exactly the same in the females' brains as it does in males.

This research focuses on the variation in allele frequencies for MAOA and the relationship that its expression has to early social and physiological experiences during development and its variation in functional outcomes in different social contexts. In other words, this is an underlying genetic element that plays an interesting role in affecting the brain structures that are associated with the expression of aggression. But the behavioral outcomes of gene variation are totally dependent on the patterns in early life experience and the social context in which some carriers of the low-production allele find themselves.[57] The bottom line is that *if* you have a low-expression allele AND you undergo severe childhood trauma or abuse, then the likelihood of your having problems in the neurological infrastructure of aggression that result in higher aggression is higher than if you had the regular-production allele.

5-HT/Serotonin

Most people have heard of serotonin, but most do not know that it is tied to the neurotransmitter 5-Hydroxytryptamine (5-HT for short).[58] Of all the well-known neurotransmitters this is the one best recognized for affecting behavior, especially aggression. The main way in which 5-HT relates to aggression has been determined from studies of the variation in the receptors that interact with 5-HT. There are at least thirteen types of 5-HT receptors and multiple molecules that interact with, and regulate, 5-HT in the brain. In general it appears that serotonin concentrations in the brain, and the way they interact with receptors, can modulate aggression and violent behavior in mammals, including humans (although the vast majority of the research has been done with rodents).

In general low 5-HT levels are associated with higher levels of aggressive or impulsive behavior and high 5-HT levels and/or manipulation of the 5-HT receptors in different parts of the brain can act to reduce aggression. Genetic evidence for these impacts comes almost completely from studies of rodents: mice or rats missing specific genes that affect 5-HT concentrations or the production of 5-HT in certain brain regions are more aggressive. But this is not true for all 5-HT, as manipulation of certain receptors and of concentrations of 5-HT in different parts of the brain have different types of impacts on aggression, anxiety-related behavior, and impulsivity. In the few studies of humans that track 5-HT relating to aggression there is a negative relationship between the ability of 5-HT receptors to bind to neurotransmitters and aggression, which supports the notion that this molecule impacts the expression of aggressive behavior.

One of the most interesting findings from the 5-HT studies is that, in rodents, different types of 5-HT receptors have different impacts on expressed aggression, depending on whether the rodents were exhibiting "normal" territorial aggression or impulsive pathological aggression stimulated by drugs or electric shock. This suggests that the various genes that code for the different 5-HT types are involved in different systems in aggression and that they might in fact have multiple, even mutually negating, roles in the production and expression of aggressive behavior depending on the social context and the type of aggression expressed. So while 5-HT is definitely involved in the expression, and probably the modulation, of the level of aggression, there is no evidence that this is where aggression comes from.

Testosterone

One cannot talk about the biology of human aggression without saying at least a few words about testosterone. There is a strong popular assumption that testosterone stimulates or enhances aggression, especially in males. First off, it is important to note that testosterone courses through both male and female bodies, but that on average, males have higher circulating levels than females. Testosterone is a steroid hormone closely related to estrogen and a suite of hormones called androgens. Little is known about the underlying genetic structures that influence testosterone, but there is no doubt that some genetic variation influences the production and regulation of testosterone in human bodies.

The concept that testosterone produced by males facilitates and increases aggression is an oversimplification. There are very weak or inconsistent correlations between testosterone levels and aggression in adult humans.[59] Even when external sources of testosterone are administered to adults their aggression does not tend to increase, nor is there an increase in aggression at puberty when human males undergo a significant increase in the production of testosterone and the development of male secondary sexual characteristics.

There is evidence that in competitive or acute stress situations humans can respond by increasing the production of testosterone but there is no strong or consistent evidence that these increases result in increased aggressive behavior. The increase does appear to enhance muscle activity and efficiency and might also result in lower sensitivity to pain or punishment in both men and women. This might make individuals more likely to participate in aggressive competition, but it does not increase aggression per se. In some experiments the levels of circulating testosterone increase after dominance interactions and social competition, but again this is not necessarily tied to overly aggressive behavior. Exposure to sexual situations and to communal competitive events (like team sports) also appears to increase testosterone in males. Interestingly, males who are fathers and or long-term married partners show lower levels of testosterone than do nonfathers or unmarried males.[60]

Overall, testosterone seems to be associated with the efficiency and activity of a variety of muscular and other physical systems, some of which are implicated in the expression of aggression. But contrary to popular misconceptions, testosterone itself is not associated in any causal way with increased aggressive behavior or in the patterns of the exhibition of aggression.

The nature of human aggression is not found in our genes

Despite popular notions that certain genes or genetic elements control or regulate the appearance and intensity of aggressive behaviors, there is no evidence for any one-to-one genetic controls, nor is there evidence for certain molecules or systems in the body that predetermine aggressive outcomes. None of the information available on the neurological infrastructures for aggressive behavior implicates these systems as uniquely responsible for aggression, especially because all of these systems have many other functions. Even in detailed studies of the molecules that directly impact portions of the brain that are implicated in the expression of aggressive behavior it turns out that the context-specific, socially mediated, and contingent nature of the expression of aggression in humans remains the best explanation. There is no gene or system in the body that can be identified as "for aggression." While it appears clear that genetic variation in neurotransmitters and hormones can be involved in the ways in which we express aggressive behavior, there is no direct or casual link. Our genes cannot make us aggressive. The myth that the nature of human aggression can be found in our genes is busted.

MYTH BUSTING: AGGRESSION IS NOT THE KEY TO "SURVIVAL OF THE FITTEST," AND BEING AGGRESSIVE AND VIOLENT, ESPECIALLY IF YOU ARE A MALE, MAY NOT GET YOU MORE BENEFITS

The myth that being more aggressive makes you more dominant, which in turn gets you the girl (if you are a guy), is quite pervasive.[61] The basis, in biology, for this myth is threefold:

1. Because sperm are cheap for males but eggs and infant care are expensive for females, the sexes have different evolutionary priorities when it comes to mating; males want to get to as many females as possible and females want the highest quality males.[62]

2. Because males tend to be larger than females they can use force to get resources and coerce females to mate.

3. If more aggressive males get more resources, females should prefer to mate with them because their offspring will also do better.

In short, males and females have different biologies, resulting in males' ability to use aggression to get more goods, and thus females should prefer the more aggressive males. These patterns are reported to hold up well for a number of insect species.[63] Some researchers also assert that they are central to the evolution and behavior of humans and our primate relatives.[64] To tackle this myth we can ask the question, does the aggressive guy get more goods and the girls? The short answer is no, aggressive males do not get more in an evolutionary sense than other males, but as always the actual answer is more complicated than simply yes or no.

To assess the assertion we need to ask the comparative question, Did this pattern evolve? by looking at our closest relatives, the primates, and then ask the current behavioral question, Is this what happens in human societies? Summarizing the behavioral datasets for primates and looking at humans across cultures demonstrate that aggression and violence do not correlate, on average and over time, with any increased success with potential mates.

Aggressive primate males do not necessarily get more females and more resources

We think about lead males as alphas. When we talk about alpha males in general people associate this with a position of dominance and prestige, and imagine that this has been obtained, at least in part, through fighting and the use of aggression. In primate societies (and in social mammals, where this term comes from), "alpha" refers to the top spot on a linear dominance hierarchy. Dominance simply refers to the ability of one animal to win contests over resources consistently. That is, if you have two monkeys and put a peanut between them, the dominant individual will go right to the peanut and take it while the subordinate animal will not even try. So dominance is a relationship between individuals, not a physical or genetic characteristic of any given individual. Dominance status emerges through social interactions, including aggression, and can change over the lifetimes of individuals.

It was long held that dominance status correlated tightly with reproductive success. However, this relationship is less clear than previously thought. In primates there are many cases where males other than the dominant individuals do successfully reproduce, others where females in one group are actually reproducing with males from other groups, and even some where observed dominance in one case (say fights over food) does not translate to priority of access to females. However, in a slight majority of primate studies dominant males, while not having exclusive mating rights, do tend to father the majority of offspring born into the group while they are dominant.[65] So we can say that, on average, a male's position in the dominance hierarchy can correlate with access to females and reproductive success (number of offspring produced).

But dominance is not simply about being aggressive; it is about skillful social negotiation. High rates of aggression by male monkeys can be detrimental to their ability to stay in a social group. For example, the primatologist Steve Suomi demonstrated that some male rhesus monkeys were extra aggressive and seemed to impulsively start fights or use aggression much more than others in the group. He found that these impulsive males were mostly driven out of their group before they could breed, suggesting that aggression by itself does not result in greater access to goods or to females. Barbara Smuts also found that in some primate species females could band together to resist or avoid the aggressive attempts of males to exert dominance over them.[66] So our real question is not about aggression by itself, but rather whether aggression leads to dominance.

In most studies of dominance relations, males use a variety of means to move up (and down) the dominance hierarchy. For example, in a five-year study of long-tailed macaque monkeys in Bali, Indonesia, my colleagues and I watched three adult males become dominant in their respective groups. Interestingly, each did so via a distinct pathway. One male followed what would be considered the aggression path. He was aggressive to all the females and males in his group, constantly picking fights (and winning them) and physically harassing other group members such that they gave way whenever he was around. The second male tended to fight when attacked, but almost never started fights. Rather, he spent nearly all of his time approaching all the females in the group and having sex with them (whether they were capable of getting pregnant or not). Over time the females started to support him when other males attacked him. One other male joined forces with him, and the two of them became the favorite partners of the females, always receiving their support in group conflicts. This is the way he ended up as the dominant male in the group. Finally, the third male spent nearly all of his time grooming and in friendly behavior with males, females, and especially the young members of the group. When fights would break out he (along with a group of young males that hung around him) would almost always go over and break up the fights. This is the way he became the alpha male in his group. However, over the entire time he was the dominant male, females frequently came up to him requesting sex and he almost never acquiesced (we only saw him mate a few times over many years).

Similar variation in dominance styles, with and without a major role for aggression, is reported for a majority of primate species.[67] This pattern of complex ways to attain, maintain, and use dominance seems to be characteristic of primates. Additionally, in most primate groups males go through a period of higher dominance as prime adults and then fall to lower ranks as they age. It turns out that age, by itself, can be a major predictor of dominance rank. Additional factors in the acquisition of dominance rank are whether a male was born into a group, how long the male has been in a group, and various other demographic and life history variables.[68] Aggression is important in the acquisition and maintenance of dominance, but it is not the single most important factor, nor is it a ubiquitous facet of all dominance cases. For primates, being aggressive does not equal getting more goods or females. The comparative approach does not suggest that males have evolved aggression as a strategy for reproduction or dominance acquisition.

What about aggression in humans?

In the United States, there is a common saying that "nice guys finish last." This reflects a popular cultural assumption that, although we value "nice" (meaning nonaggressive) in males, those males who are more assertive, aggressive, and dominant will do better in society and be more attractive to women. This is pretty much the same assumption outlined above for the primates—and shown to be less than accurate. But is it true for humans?

Some would point to the common occurrence of male violence against women in modern society and say yes. The psychologists Margo Wilson and Martin Daly state that "certainly, it is easy to envision circumstances in which a man's capacity to use violence effectively might enhance his attractiveness to women . . . even where sexual harassment and assault are prevalent, a husband with a fierce reputation can be a social asset."[69] This perspective reflects an assumption that aggression toward women by men enables an evolutionary advantage for the males, including more resources and more offspring. These females might even be selecting for more aggressive men by preferring to mate with those who are more aggressive because they are the best providers or because the women (and their offspring) are protected from aggressive actions by other men. But is there evidence to suggest that this actually happens?

In fifty intensively studied hunter/gatherer societies the anthropologist Christopher Boehm notes that deviances from typical behaviors such as being overaggressive or stingy, stealing or cheating, bullying or being an unpredictable repeat killer received strong negative reactions from members of the group. This included active social sanctioning, such as direct criticism and ostracism ranging from mild to group ejection, alongside more emphatic actions such as capital punishment and supernatural sanctions.[70] Boehm also notes that rather than preference for aggression there appears to be positive reinforcement of nonaggressive traits in adult men in most of these societies. This perspective is widespread across much of the anthropological literature—not that we are peaceful by nature or that men are not aggressive, but that in the majority of human communities there is a constraint on aggression and a favoring of prosocial interactions.[71]

Wilson and Daly and their colleagues who hold the same viewpoint are largely influenced by datasets for North American domestic violence and gang violence. Indisputably, men are aggressive toward women and violence does occur in our society. But does this reflect an evolved strategy by males? Boehm and his colleagues make their arguments based on overviews of small-scale societies (foragers) and different communities across the planet. These are interesting assertions, but are there any actual behavioral and reproductive data that support the notion that aggressive men get more mates and more resources across the human species? There are extremely few studies of human groups that actually strive to test the question whether aggressive males do better. There is one extremely famous study that is held by many to support the perspective that male aggression is an evolved strategy. This study, almost always considered the final proof on the matter, is of the men of the Yanomamö, a horticulturalist tribe in South America.

The Yanomamö are put forward by some as the best example of the evolution of aggression in humans and a stellar example of how aggression and reproductive success go hand in hand for human males. However, does the story about them hold up to scrutiny?[72] The Yanomamö live in villages associated with territories and their gardens. They occasionally raid other villages, sometimes abducting females during the raid. The Yanomamö engage in relatively high rates of aggression and violence breaks out within and between villages, sometimes resulting in death. If a Yanomamö man is involved in a killing he must undergo a purification ritual signifying that he has become a unokai. Only a minority (less than 30 percent) of men become unokai (that is, actually kill someone); however, it has been claimed that unokai have, on average, 2.5 more wives and three times the number of children than do non-unokai men. This difference has become extremely famous as the most cited example of aggression and reproductive success in human males.

However, there is a problem with the assertion. While unokai do have, on average, more offspring, they are not being compared with a group of men who are of the same age. We know from primate data that age can affect dominance and reproductive success. Douglas Fry's reanalysis of the original Yanomamö dataset shows that unokai are on average 10.4 years older than non-unokai. It is not surprising that, on average, men who are ten years older have more wives and children.

This blurs the relationship between being a unokai and reproductive success. Also, Yanomamö headmen tend to have more wives and children and there is no examination of headman status and reproduction in the original reports. Adding to this investigation is the recent report on the Waorani, another well-known and much-studied South American tribal group known for their aggressiveness and with the highest rate of homicide of any society. Stephen Beckerman and colleagues examined the genealogies of 121 Waorani elders and the complete raiding histories of 85 warriors.[73] They analyzed raiding histories, marital trajectories, and reproductive histories and discovered that "more aggressive men, no matter how defined, do not acquire more wives than milder men, nor do they have more children, nor do their wives and children survive longer." In fact, they found that more aggressive men had fewer children surviving to reproductive age. In neither of these two well-known, and well-studied, examples of aggressive societies, where reproductive success was actually measured, do aggressive males do better.

More aggressive, more violent, or more warlike males do not do better among humans and our closest relatives

Males are larger and can use physical aggression more violently than females, but it does not appear in humans or other primates that these actions result regularly in evolutionary benefits for males. This is not to say that aggression is not important or that humans and other primate males do not use aggression to further their own goals. It is just that the available evidence shows that aggression is neither the primary, nor the most successful, way to achieve dominance and to mate and produce offspring. Although humans and male primates can be aggressive, and humans can be especially violent, there is no indication that this is a consistent or evolved strategy in our species. The myth that aggression is an evolved strategy and that being aggressive and violent, especially if you are a male, get you more benefits is busted.

MYTH BUSTING: HUMANS DO NOT RELY ON AGGRESSION AND VIOLENCE MORE THAN ON COOPERATION AND MUTUALISTIC INTERACTIONS

That humans are selfish and uncooperative by nature is the easiest of all the myths to bust. Humans do use aggression (sometimes quite a lot), but its frequency does not even come near the frequency of prosocial and cooperative behavior. There is a substantial body of research demonstrating that humans are more often involved in reciprocal positive interactions and a wide array of cooperative behaviors than in selfish behavior or aggression. Much as Robert Sussman and Paul Garber demonstrated for primates in general, the vast majority of daily interactions, behaviors, and relationships for humans involved prosocial and cooperative behavior rather than conflict and aggression.

There is a broad set of behavioral and evolutionary literature supporting the notion that cooperation is core to human societies, even if we consider the height of violence, human warfare. While cooperation is not synonymous with peace, it does imply social and behavioral systems that are far more complex than aggression and violence-based dominance hierarchies. It also turns out that humans cooperate at extremely high levels on a daily basis, largely without much aggression at all.

Humans are hyper-cooperators and are not generally selfish

At the heart of the uniquely human way of life is our particularly intense, mentally mediated, and highly structured way of interacting with one another . . . that itself relies both on communication and on a level of cooperation unique in the animal world.

—Nicholas Enfield and Stephen Levinson (anthropologists)[74]

. . . [there is] extraordinary cooperation between nonkin, including specialization, a regular flow of goods and services between individuals and groups, and the formation of increasingly complex alliance networks. Human cultural conventions solve the problem of reliably and stably associating cooperating individuals with each other.

—Kim Hill, Michael Barton, and Magdalena Hurtado (anthropologists)[75]

. . . we cooperate with one another widely, in relations of exchange and reciprocity, in hierarchies and as equals, in relations of production as well as distribution, in experiments on sharing, monitoring, and punishing as well as in daily life, and in institutions that intentionally foster cooperative behaviors.

—Robert C. Marshall (anthropologist)[76]

Cooperation is basically when an organism behaves, possibly along with others, in a manner that incurs some cost for the individual but benefits one or more other individuals (potentially including the one acting cooperatively). It appears that cooperation and shared information exchange, combined with socially negotiated distribution of labor, seem to effectively coordinate large groups of people: simply put, cooperation is what humans do best and what makes us such a successful species.[77] This point is not new; it has been reiterated, researched, supported, and repeatedly demonstrated since the time of Darwin. The quotes above are drawn from a small subsection of the diverse literature on human cooperation, all of which shows that humans are unique relative to other forms of life on this planet in the extent and complexity of their cooperative patterns.

The case for cooperation as central to human success has been made for human ancestors, modern humans in simple foraging societies, agriculturalists, and modern nation states. There is copious fossil and material evidence that from early on in our evolutionary history the ways in which we worked together as opposed to selfish and individually based behaviors is what enabled humans to spread far and wide across the planet. There is also a large body of evidence that demonstrates that agriculture, village and city structures, large-scale religious interactions and political systems, and trade and market economies all rely on a substantial infrastructure of human cooperation for their success.[78] This does not mean that competition and conflict are not also common, just that these are not the basis for our success. Aggression can emerge out of cooperation, or the breakdown of cooperation, but nearly every study conducted on human social behavior indicates higher frequencies and greater emphasis on cooperation than any other single behavioral pattern.

This prevalence of cooperation does not mean that individuals do not act selfishly on occasion or that some might use aggression to take from others. It means, contrary to some perspectives, that a primarily selfish orientation (sometimes referred to as *Homo economicus*) is not characteristic of most people in most societies. This basic notion of *Homo economicus* is that humans as individuals will make decisions based on what is best for themselves. This turns out not really to be the case. In a study of fifteen societies, the research team led by Joseph Henrich and colleagues demonstrated unequivocally that the central axiom of *Homo economicus* (humans will behave selfishly in economic situations) is refuted. In fact, selfishness as a primary pattern was not found in any of the societies studied. Rather, patterns of cooperation and social reciprocity were dominant, with much variation in details across societies, based on integration into world markets as well as demographic and other social variables.[79] This does not mean that humans are all egalitarian or that we are selfless. It simply reflects the reality that human societies are based on extensive and extremely complex systems of cooperation and mutual inter-reliance on one another, such that a consistently selfish behavioral strategy will not be sustainable in human groups.

This prevalence of cooperation does not negate aggression or violence and in fact probably enables the kind of intense and extreme violence that is characteristic of modern warfare and civil conflicts. To create and maintain armies you need extremely complex cooperation and to engage in wide-scale warfare, coordination and a near complete suppression of selfish behavior is needed. One might even argue that war is possible directly due to humans' unique abilities to maintain large-scale and intensive cooperation.

While cooperation is not the opposite of aggression, its origin and expression reflect a different underlying set of goals and behavior. As humans we do get along pretty well (most of the time, without using aggression) and cooperation is at the heart of how we do it. That humans are ultra-cooperators and not naturally selfish tells us that as a species we do not rely on aggression and violence more than cooperation. There is no pattern of evidence to support a notion that humanity is aggressive and selfish by nature: this myth is busted.

Human behavior can be amazingly violent, but violence is not intrinsically human

It is obvious that human aggression is an amazingly complicated thing. There is variation in conflict styles and aggression across individuals, sexes, genders, societies, and time frames. Aggression is an important part of being human, but it is not who we are at our core.

We now know that aggression itself is not a uniform or consistent discrete trait, so aggression per se cannot be favored by evolutionary pressures to form the basis of the human experience. The other primates show us that we do not have specific, evolved patterns of heightened aggression, especially in males. Looking at the chimpanzee species demonstrates the potential for variability in the expression of aggressive and nonaggressive behaviors in our shared ancestors. War is common in the human experience today, but it is not a central part of our evolutionary heritage.

We know that males and females differ in some facets of aggression, but a lot of those differences have to do with physical size and the social and experiential contexts in which they find themselves. We know that more aggressive, more violent, or more warlike males do not necessarily do better, either in humans or in our closest relatives. Human aggression, especially in males, is not an evolutionary adaptation: we are not aggressive, big-brained apes.

We know the regions of the brain and body that influence normal aggression. While our genes do not control or determine the normative expression of aggression, abnormal biological function can influence particular patterns of aggressive behavior. The nature of human aggression is not found in our genes, but understanding the function and variation in our biology can help us better understand the pathways and patterns of aggressive behavior.

As a species we do not rely on aggression and violence more than cooperation; there is no pattern of evidence to support a notion that humanity is aggressive and selfish by nature. The myth of a human nature characterized by an intrinsic aggressiveness is simply not true.

MOVING BEYOND THE MYTH

If we weigh the totality of the evidence, we arrive at a new conclusion: Humans are not really so nasty after all.
—Douglas Fry (anthropologist)[80]

We will never understand violence by looking only at the genes or brains of violent people. Violence is a social and political problem, not just a biological and psychological one.
—Stephen Pinker (psychologist)[81]

When the anthropologist Richard Wrangham was asked the question, "Are humans predisposed to behave violently?" he answered, "Well, to talk about inherent aggression in us sets off alarm bells for some people because it sounds biologically determinist, it sounds pessimistic. So, I wouldn't want to quite put it in that term. But, I do think that there's all sorts of evidence that humans have got a predisposition to behave with violence in certain contexts."[82] In spite of my using quotes from Wrangham's book *Demonic Males* to exemplify the perspective of an evolutionary core to human violence, it is apparent that, along with most researchers, Wrangham does see that context and variability are central to any sincere attempt to understand human violence. The Dutch philosopher Raymond Corby notes that over three hundred years ago the political philosopher Thomas Hobbes proposed that social order (cooperative society) is not intrinsic to human nature, but is installed by a social contract that constrains and pacifies the natural violent state of humankind and the solitary individual's brutish natural tendencies. Hobbes's assumption, that humans have an inherent violent nature which must be modified by society, is reflected in many of the perspectives discussed in this chapter. However, even when there is attention given to context and variability there is resiliency to the popular myth about human nature and human beings' predisposition to aggression.

The information reviewed in this chapter demonstrates that such a perspective is incomplete and paints too simplistic picture of how and why human beings behave aggressively. Sure, certain things spur aggressive actions, but the common notions about our inner, natural, aggressive tendencies (especially in males) ignores the complexity of human biology, psychology, history, and society. It downplays the myriad ways in which aggression is initiated and maintained, and oversimplifies what we can mean by, and understand about, human aggressive behavior. And, most dangerously, it enables a kind of inevitability in our communal sense of aggression and society, especially as it relates to males. This need not be the case.

As the anthropologist Ashley Montagu sagely cautioned, "It is essential that we not base our image of ourselves on false foundations. What is involved here is not simply the understanding of the nature of humanity, but also the image of humanity that grows out of that understanding."[83] Humans are not naturally aggressive, but they do have a great potential for aggression and violence. If we believe we are aggressive at our base, that males stripped of social constraints will resort to a brutish nature, then we will expect and accept certain types of violence as inevitable. This means that instead of really trying to understand and rectify the horrific and complex realities of rape, genocide, civil war, and torture, we will chalk at least a part of these events up to human nature. This is a dangerous state of mind that traps us in a vicious cycle of inaction and futility when it comes to moving forward as societies invested in understanding and managing violence.

Understanding humanity is neither simple nor linear; as humans we are neither violent nor peaceful at heart. We can be aggressive and violent and peaceful and cooperative all at the same time; arguing for a natural state of peace or a natural state of aggression is missing the boat. However, in this chapter we have taken on the myth of a natural aggression and demonstrated that it does not hold up in light of available evidence. What can we make of this? If we are not naturally aggressive how do we explain and understand human aggression and violence today? We do so by opening our minds to the vast amount of research across the social and biological sciences, to histories and critical assessments of social patterns and lives. We also commit ourselves not to accept simplistic assertions about who we are and what we are capable of. We strive to use this information to make educated and socially involved decisions about aggression and violence. In the end, we try to keep an open mind and cooperate with one another to understand and resolve the impacts of aggression and conflict.

6

Myths about Sex

In 1995 the author and family therapist John Gray published the first edition of his book *Men Are from Mars, Women Are from Venus,* which argues that to make male-female romantic relationships (especially marriages) work, one needs to realize the core differences in communication, emotion, and behavioral styles of males and females. Twenty years (and multiple editions and follow-ups) later, this is still a common metaphor people use to think about men and women.[1] Men are aggressive, belligerent, but protectors like the Roman god of war Mars, and women are emotive, beautiful, vain, and fertile like the goddess of love Venus. This implies that males and females want to have a specific kind of romantic relationship, but that males and females speak different languages, have different desires and needs, and although they are the same species, act like they come from different planets.

In our daily lives we are constantly bombarded with images, words, and situations that reinforce the notion that men and women differ in bodies, desires, needs, and even minds. A book entitled *The Teenagers Guide to the Real World* starts its chapter 11 with a phrase that summarizes the myth: "Men and women are completely different." The book goes on with an exaggerated, but not unfamiliar, explanation for why and how the sexes differ:

> Men are equipped to impregnate women. There is no cost to a man in impregnating someone. Women, on the other hand, are equipped to be

impregnated and produce babies. As soon as a woman gets pregnant she has just signed on for a 20 year tour of duty taking care of the resulting child. Her goal, going back millions of years, is to help that baby survive. For a woman pregnancy carries an extremely high cost. Furthermore, the woman's mind and body also know, instinctively at some level, that a baby needs two people to survive. Women are therefore designed to wait for a strong commitment prior to getting pregnant. In our culture that commitment is called "marriage," and women are smart to wait for it. Many men seem to have little or no such programming. This basic anatomical difference, by itself, leads to rather strong differences in priorities between men and women. In addition, men and women clearly have different programming in other parts of their brains.[2]

These books reflect the common perception that men and women "complete" each other in their differences, that marriage and the quest for a perfect mate emerges from our evolutionary histories, and that male aggression and female nurturing are part of the package. Although a bit over the top, the preceding quote highlights the point that our perception of male-female differences relies heavily on current popular beliefs about the mind, the body, and evolution: it is widely accepted that male and female differences are a reflection of our nature.

MEN AND WOMEN ARE FROM DIFFERENT PLANETS, AREN'T THEY?

Most people seem to think so. It is a common assumption that parts of the male and female brain have evolved to focus on different things; men want sex and sports, and women want material things, to be social with other women, and avoid sexual advances of men. A core part of these differences is sexuality: it is a basic assumption that males and females see sexual activity in very different ways. This view (and its association to the overall myth) is evident across many aspects of our culture. Think of the time leading up to Christmas and Valentine's Day when the media is packed with advertisements for jewelry, always showing the man buying a diamond for the woman, and the woman being eternally grateful; this image of gift-giving is a metaphor for men providing goods or support in exchange for women giving them access to sex (or a bond of marriage and its association with continuous sexual access). Think of the advertisements for online dating sites, which focus on the cultural goal to marry, its relation to sex and sexuality, and the concept that there is someone out there for everyone. There is near total agreement that at the heart of it men and women want different things out of life and sex, as the journalist Nicholas Wade asserts: "When it comes to the matter of desire, evolution leaves little to chance. Human sexual behavior is not a free-form performance, biologists are finding, but is guided at every turn by genetic programs."[3]

The concept that there is a well-established pattern of differences between the sexes is a belief about human nature. But is this belief justified? What if sex and sexuality are really complicated? What if our assumptions about what is normal and natural are not reflected in the actual data about sex differences and similarities? The following two quotes challenge the myth about patterns of human sexual differences by suggesting that male and female behavior might not be so different or that differences might not be as ingrained as we think they are:

> Although sex is a biological urge, it is rarely experienced in the same ways by people everywhere: it is differently practiced and felt depending on the social and cultural settings in which it occurs. (Hastings Donnan and Fiona Magowan, anthropologists)[4]

> The gender similarities hypothesis holds that males and females are similar on most, but not all, psychological variables. That is, men and women, as well as boys and girls, are more alike than they are different. . . . Results from a review of 46 meta-analyses support the gender similarities hypothesis. Gender differences can vary substantially in magnitude at different ages and depend on the context in which measurement occurs. . . . The question of the magnitude of psychological gender differences is more than just an academic concern. There are serious costs of overinflated claims of gender differences. These costs occur in many areas, including work, parenting, and relationships. (Janet Shibley Hyde, psychologist)[5]

Hyde also suggests that if these assumptions about human sexual differences are incorrect, their maintenance might even be detrimental to our society's functioning and health. How do males and females actually behave? Do our cultural schemata filter how we see and interpret the world or are the differences we seem to see in everyday life accurate representations of a human nature?

As with the myth busting in the previous two chapters, reality is not simple but it is important. Sex and sexuality are very complicated and they mean a lot for our daily lives. What we really know about men and women and the nature of sex in humans challenges the extent of these differences and any simplistic take on this topic. To bust this myth we have to test the core assumptions and refute them.

Testing Core Assumptions about Sex

ASSUMPTION: *Males and females are biological very different from one another.*

TEST: Are male and female biologies totally different, sufficiently different, or just versions of the same biological theme? If there is a clearly distinct biological patterning between males and females that mandates radical differences in behavior and function then the assumption is supported; if males and females are basically variations on a theme, and not that different, then it is refuted.

ASSUMPTION: *Behavioral differences between males and females are evolutionary; they are hardwired.*

TEST: If the differences in behavior between males and females are more biologically based (sex) than culturally based (gender) and are best explained as evolutionary adaptations, the assumption is supported. If, however, the differences are complicated, less clear, and mostly related to patterned social differences between genders, not primarily to evolved differences, then this assumption is refuted.

ASSUMPTION: *Males and females are different because they are complementary to one another, resulting in the monogamous pair bond and the nuclear family as a natural state for humans. This means that it is a natural human goal to obtain a unique and powerful sexually monogamous romantic relationship.*

TEST: This has a multipart test: first, are humans monogamous sexually? If yes, then supported; if no, then refuted. Second, are pair bonds and marriage (or at least romantic relationships) the same thing? If yes, then supported; if no, then refuted. Finally, do humans "naturally" live in nuclear families where the strongest bonds are between husband and wife and children? If yes, then supported; if no, then refuted.

ASSUMPTION: *Men and women are really different when it comes to sexuality: men want sex and women want relationships (and less sex than men).*

TEST: Do men want more sex than women? Are men more sexually focused than women? Do the sexes differ dramatically in how, when, and how much they have sex? If yes, then supported; if no or if it is much more complicated than these simplistic assumptions, then refuted.

MYTH BUSTING: MALES AND FEMALES ARE MADE OF THE SAME BIOLOGICAL STUFF

Are male and female biologies totally different, sufficiently different, or just versions of the same biological theme? This section of the chapter summarizes what is known about the development of humans into male and female sexes and the differences and similarities between adults. From the development of the male and female reproductive tracts to the range of variation in sex chromosome patterns, to the physiological, morphological, and neurological variation and overlap of human sexes the bottom line is, while there are many differences, there is no doubt that we all are the same species and are more biologically similar than different.

Of course, no one in their right mind is going to deny that there are differences from birth (or even before) between males and females:

> Yes, boys and girls are different. They have different interests, activity levels, sensory thresholds, physical strengths, emotional reactions, relational styles, attention spans and intellectual aptitudes. The differences are not huge and, in many cases, are far smaller than the gaps that separate adult men and women. (Lise Eliot, neuroscientist)[6]

However, those differences are not necessarily what we think they are, nor is the gap as wide as is usually presented. In fact, in many cases there is no gap at all. One concept critical to our discussion needs to be examined prior to reviewing the biological and behavioral data—overlap of distributions. When we talk about differences we tend to think of a point on a line or single figures, not the entire range of variation that actually occurs. For example, we already mentioned the size dimorphism in our species, with males 10 to 15 percent larger than females. These percentages represent an average difference, with both males and females showing a large range of variation with substantial overlap. In figure 6 we see one male who is about 12 percent taller than the female. In figure 7 we can see the total range of male height and the total range of female height with the means separated by about

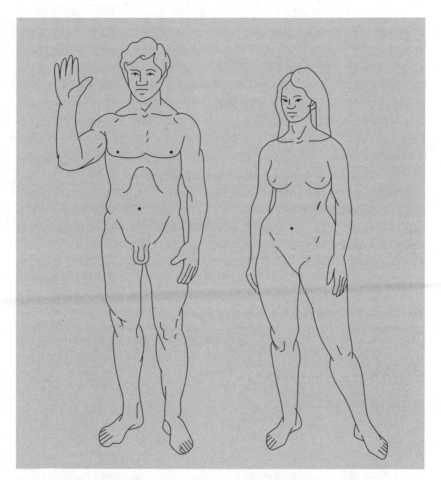

FIGURE 6. In this illustration sent out beyond the solar system on the *Pioneer 10* spacecraft, the representative human male is about 10 to 15 percent larger than the female. Adapted from NASA.

12 percent, but notice the substantial overlap. In practice, when we use only averages we ignore the actual real-world patterns where there is a lot of overlap. So when we say males are 10 to 15 percent larger than females we don't mean that every male is larger than every female, just that the averages in height between the two groups are separated by that figure. Remember this as we discuss the differences (and lack thereof) in males and females; sometimes it is important to see the forest and not just a few tress to understand what is really out there biologically and what is the product of culturally filtered schemata.

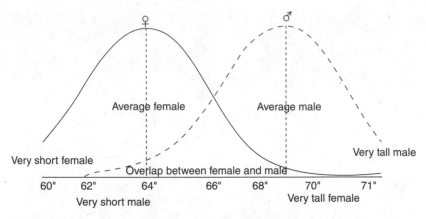

FIGURE 7. Total ranges of male and female heights, with the means separated by about 12 percent. Notice the substantial overlap.

Basic Physical and Developmental Patterns

The basic core of the physical difference (and the legal definition) of male and female does reside in our DNA or, more specifically, in our chromosomes. Generally, we call someone with two X chromosomes a female and someone with one X and one Y a male. If you have one X and one Y there are genes on the Y chromosome that initiate the development of male physical patterns. If you have two X's then the active genes initiate the development of female physical patterns. These developmental trajectories include different patterns of hormonal action, muscle and bone development and, possibly, some brain differentiation.

As usual, nothing is as clear as we'd like it to be. The XX and XY classification does not always correlate with the physical and behavioral patterns we associate with male and female. You can be XY and have an error in the activation of segments of DNA so that the specific genes that initiate male development (like the genes called TDF and SRY) never turn on and their protein products are not properly made or transported. In this case the genetic impact from the one X will facilitate the development of female physical form. There are also a wide array of other variations on this theme such as XO (no Y), XXY, and a variety of developmental scenarios which cause less than crystal clear sex outcomes such as XX individuals with male genitals,

XY individuals with female genitals, individuals who are XX or XY but have mixed sex genitals, and other variants.[7] The total frequency of variations on the standard patterns (XX equals female physical form and XY equals male physical form), which is called intersexed, is about 1.7 in every 1000 births. For a little context on this, the average frequency of albinism is about 1 in 20,000 births. Intersexed individuals, biological variation that muddles the clear distinction between what can be genetically defined as male and female, are fairly common in humans. Most of the cases are minor in effect and the individuals are able to generally confirm to the physical expectations of one sex or the other, but this still reflects a pretty flexible system of sex development.

All humans, male or female, share the same bones and physical structures. We are the same species and all of our tissues (such as tendons, ligaments, bones, blood, skin, etc.) are made of the same stuff. But these bodily tissues do not always take the exact same shape. As was hinted at in chapter 5, there are a suite of physical differences that, on average, occur between males and females. A pronounced difference is found in the shape of the pelvic bones. In females, the pelvic girdle is more flared outward and the size of the central space created by the bones of the pelvis (the birth canal) is larger in females than in males. The reason behind this physical difference is obvious: females give birth and thus need maximum space for the birth canal. It is this larger birth canal and the wider flare of the pelvis in females that gives them slightly more side-to-side displacement when they walk than males (on average). This behavior, the slight swinging of hips when walking, is often accentuated culturally to reflect a hyper-feminine behavior.[8] Here is our first example of a small physical difference between males and females that can be culturally played up to be visually very distinct.

Aside from the pelvis, there are other common patterns of physical differences. In general female skulls have a more vertical forehead, smaller ridges above the eyes, fewer bony buildups around muscle attachments, and smaller mastoids (the lump on your skull just behind and below your ears). Also, the angle of the female jawbone tends to be larger than in males. This gives an overall rounder and smoother look to females heads than those of males. In addition to these skeletal differences males tend to have, on average, higher muscle density per unit area and more upper body strength than females (which contributes to their greater physical impact in aggressive behavior, as discussed in chapter 5).

The way we deposit fat on our bodies contributes a great deal to their shape. Human females and males lay down fat (called adipose tissue) more or less in the same way that most mammals do, but at slightly higher rates. That is, modern humans tend to be a bit fattier than most mammals. The pattern of fat deposition is similar for both sexes in location, but males and females exhibit slightly different rates of fat deposition by location and of fat utilization: females deposit more fat around the chest area and thighs, men more around the abdomen, and men burn or utilize fat faster than women.[9] Because we walk upright on two legs (and have associated changes in our muscles and their attachments), and the fact that we only have two nipples (many mammals have multiple sets), the pull of gravity on that fat and the structure of our bodies create different fatty accumulations than in other mammals (like breasts for human females and pronounced buttocks for both male and female humans relative to other primates). This also creates the differences in the general appearance of male and female human bodies.

Remember that all of these variables are average patterns. In any given population you will find some males smaller than some females and some females whose skulls and bodies have many male characteristics, and vice versa. The patterns of difference between male and females bodies are there, but any specific individual will have some variation on these themes. Also, when you compare between populations or across the whole species these patterns of difference are less robust (there is more overlap between males and females) because of the dramatic size and shape plasticity in modern *Homo sapiens* (remember the discussion of human variation in chapter 4).

We do expect substantial differences between males and females when it comes to reproductive biology, and indeed the reproductive organs and external genitalia do differ in certain ways. However, these differences are more of degree than kind, since the same embryonic tissue mass gives rise to both female and male genitalia. Females, and not males, give birth and produce milk for the offspring (called lactation). Because of this females also have specific differences in the internal structures surrounding the nipples that men lack (called mammary, or milk, glands). The tissue mass of the chest is from the same embryonic structures so both males and females have nipples that are generally the same (surrounded by a suite of glands and a cluster of nerves). But males are not able to lactate because their glands (largely sebaceous, or sweat, glands) do not develop into mammary glands. The male reproductive tract is active (making sperm, called spermatogenesis) for the majority of men's lives, with the effectiveness of the system dropping off as they age. Females, however, cease reproductive cycling and undergo a change in their hormonal patterns at some point between forty-five and sixty years of age, called menopause, which signals a cessation of their reproductive lifetime.[10] Throughout the human life span all the actual hormones in males and females are the same (there are no male-only or female-only hormones), but there can be differences in the levels, patterns, and impacts of some of these hormones in male and female bodies.

Genitals are surprising

Almost everyone assumes that the best way to tell someone's sex is by looking at their genitals: if you have a penis, you're male, or if you have a vagina, you're female. Most people think that male and female genitals are about as different as can be. But this is not true. Men and women are all made of the same stuff, even our genitals. We are just different variants on common themes.

The female and male reproductive systems, including the genitals, emerge from the same mass of embryonic tissue. For the first six weeks of life the specific pre-sexed tissue masses develop identically. Starting at about six to seven weeks after fertilization, depending on whether the fetus has XX or XY chromosomes (usually), a series of hormone and other chemical signals are distributed to these tissues and they begin to differentiate. One part of the tissues begins to form the clitoris or penis (depending on the chemical cues) and another forms the labia or scrotum. Another area begins to form into either the testes or the ovaries.[11] This means that physiologically, male and female genitals are made of the same stuff and work in more or less the same ways. For example, in sexual response physiology the clitoris and penis are basically the same. This also helps us understand the large range of variation in genitals. Male penis size and shape vary, female clitoris and labia vary, and there can even be a fairly high rate of genitals which develop without a 100 percent clear distinction between being a so-called male or female structure.[12] In large part all of this variation is irrelevant to reproductive function; the vast majority of genitals work fine. Genitals can be both an indicator of difference between men and women and a clear sign of how similar we are.

Biology of Sex and Reproduction

The core of the biology of sex is the reproductive system. Even though the systems are derived from the same tissues via similar processes, there are some important differences in the final forms. The female reproductive system includes the external vagina and clitoris and the internal uterus, ovaries, and fallopian tubes. The ovaries are involved in the storage of egg cells and the production of many hormones, such as estradiol and progesterone (also produced by males, but from different sources). Once females pass puberty they begin to cycle reproductively. Once per cycle an egg is transported to the uterus via the fallopian tubes, where contact with sperm and fertilization can occur. If an egg is fertilized it will be implanted in the uterine wall and begin development. The uterus changes its internal structure on a regular cycle (the menstrual cycle), which is tied to the timing of egg release and potential fertilization. If an egg implants, the uterine lining begins the pregnancy cycle; if not the lining sloughs off and another cycle is begun. This cycle is regulated largely by hormones such as follicle-stimulating hormone (FSH) and luteinizing hormone (LH). Females also have mammary glands, which produce a highly nutritious supplement for the newborn infant in the form of milk. Most mammals have six or more mammary glands, but most primates (monkeys, apes, and humans) have only two. The amount of fatty tissue around the glands is larger in humans than in most primates (resulting in breasts). The development of breasts at puberty is one of the secondary sexual characteristics that are regulated by the hormones secreted by the ovaries.

The male reproductive system consists of the external penis and scrotum, which contains the testes and the epididymus. Sperm are produced in the testes, transported across the accessory sex organs, and eventually ejaculated via the penis. The vas deferens (a tube-like structure) connects the testes to the seminal vesicles and the Cowper's and prostate glands, all of which are involved in the production and ejaculation of sperm. Unlike the eggs, sperm can move on their own, and once deposited in the vagina they attempt to move up into the fallopian tubes and contact an egg. As in females, the male testes (counterpart to the female ovaries) are important in the production of hormones, such as testosterone. It should be no surprise that the development and function of the male reproductive tract is heavily regulated by the same hormones that regulate much of the female reproductive function, including follicle-stimulating hormone (FSH) and luteinizing hormone (LH).

A key facet of the reproductive system is the ability to actually have sex. Not surprisingly, given the similarities between males and females, there is a high degree of overlap in how these systems function during sexual activity. Both males and females require the limbic system (the basic emotive and nervous system) to be stimulated by sensory input to initiate sexual arousal. There is some evidence that females are more susceptible to smells and possibly even pheromones than are males, and that males are more stimulated via visual cues than females. Once the initial stimulus is engaged, neural stimuli (brain actions) are transmitted via the endocrine system (using hormones like testosterone, vasopressin, and oxytocin in both males and females). This leads to physical and psychological excitement that includes increased blood flow and swelling of tissues (called vasocongestion) and a tensing of the muscles (called myotonia) throughout body, sporadic increases in blood pressure, lubrication in the vagina and inner labia (females), erection of the clitoris (females), and erection of the penis (males), glandular secretions across parts of the body, and ultimately (or potentially) a variable suite of physiological changes associated with orgasm, including male ejaculation. This is simply the physical description of sexual activity. At the end of this section we go into more detail about the wide range of variation in sexual behavior in humans and the similarities and differences between men and women.

The reproductive system and the evolution of male-female differences

One of the core assumptions about the behavioral differences between the sexes comes from a basic misunderstanding of the mammalian reproductive system. According to the *Teenager's Guide to the Real World,* "Men are equipped to impregnate women. There is no cost to a man in impregnating someone. Women, on the other hand, are equipped to be impregnated and produce babies." The basic idea is that for men reproduction is cheap and for women it is very costly. This is derived from the basic notion of anisogamy (different-sized sex cells—big egg and small sperm). Basically, the assumption is that females spend a lot of energy to produce a limited amount of eggs and then make a high investment in rearing the young, whereas males produce lots and lots of cheap sperm only. Many biologists have argued that because of this differential in the evolutionary cost of reproduction males and females should have very different approaches to reproductive behavior. Thus a male should try to fertilize as many females as possible and females should be extremely choosy and pick only males that either will help them raise the offspring or will provide the best possible set of genes for the offspring (or both). Drawing on a series of perspectives since Darwin's time the theoretical biologist and biological anthropologist Robert Trivers (and subsequently many others) translated this basic idea into evolutionary models that offered behavioral predictions for many organisms, including humans. The basic idea is that distinct reproductive pressures on males and females result in competition between the sexes caused by the differential goals and patterns.[13] This is the notion that is the basis for the belief about differences, and disagreements, between the sexes.

Not everyone agrees. The biological anthropologists Monique Borgerhoff Mulder and Kristin Rauch in their recent evolutionary overview of sexual conflict in humans point out the problem with this myth that male reproductive investment is cheap: "And as our review has shown, predicted outcomes are shaped by many factors other than sex differences in postzygotic investment in offspring. . . . More fundamentally, of course, the identification of winners and losers is a flawed pursuit."[14] Males cannot simply walk up to females and inseminate them. In social organisms, especially complex ones like humans, mating is part of a larger social reality and thus the behavior, the costs, and the contexts of reproduction are tied to a variety of factors, not just eggs and sperm. Even at the level of sperm, not just one is need for a successful copulation but rather millions (per ejaculation), so sperm are nowhere near as cheap to produce as we are led to believe. Recent work by the biological anthropologist Sarah Hrdy, et al. has also demonstrated that humans have been cooperative breeders for a long time. At an early stage in our evolutionary history multiple individuals (females and males) were involved in raising and caring for children.[15] The idea that it is natural for one human female to raise her offspring alone, or with just a single male, is a very recent one indeed, and one that is biologically not supported.

The notion that male and female behavioral differences are largely explained by the differences in their reproductive biology is absurdly oversimplified. There is a wide range of recent reviews and refutations of this notion, suggesting that a real understanding of reproductive systems and patterns of investment, aspects of sexual selection, the division of labor, and the wide array of human ecological, social, and historical contexts better explain male and female reproductive relationships than overhyped differences in their respective reproductive investments.[16]

Sex and the Brain

. . . what I found after an exhaustive search, was surprisingly little solid evidence of sex differences in children's brains. Sure, there are studies that *do* find differences, but when I looked closely at *all* the data—not just the research that confirms what we already know about boys' and girls' behavior but a truly balanced collection of findings—I had to admit that only two facts have been reliably proven: boys' brains are larger than girls and girls' brains finish growing earlier than boys'.

—Lise Eliot (neuroscientist)[17]

Males have bigger bodies and bigger brains, on average, than do females. Because of the assumptions about how males and females differ in behavior, there has been an intensive search for measurable biological differences in men's and women's brains. Over the past fifty years or so there have been many studies of the brains of cadavers and in the last few decades researchers have been able to move to various imaging technologies to examine the brains of living individuals. Yet as Lise Eliot observes, the end result of these studies does not provide any clear pattern or indication of differences that can be tied to behavior and/or other male-female distinctions. However, there are some areas of interest in the brains of males and females that have been the focus of these inquiries.

In chapter 5 we noted that the prefrontal cortex of the brain was an important region for aggression. It should not be surprising that another area near this region, the ventral frontal cortex, is of interest in studying behavioral differences. The ventral frontal cortex consists of the orbitofrontal cortex and straight gyrus, and plays an important role in normal social behavior in humans. Specifically, this area is assumed to play a role in social perception (figuring out social scenarios and contexts). In two small studies the ventral frontal cortex was shown to be slightly larger in females, suggesting that this might be correlated with females' (presumably) more acute sense of social interactions. A later study found that there was no difference in the orbitofrontal cortex between males and females, but that the straight gyrus was proportionally larger in women. This same study also correlated the larger size of the straight gyrus with "higher identification with feminine characteristics and better performance on a test of social cognition."[18] This suggested that maybe the straight gyrus had some association with female behavior and might be a good place to look for the male-female brain differences.[19]

In a study of seventy-four boys and girls the neuropsychiatrist Jessica Woods and colleagues found no pattern of differences between boys and girls in the ventral frontal cortex or the orbitofrontal cortex, but did find one in the straight gyrus: it was larger in the boys. This was the opposite result from previous studies. However, there was a twist—this size difference was negatively correlated with age. That is, the older boys had smaller straight gyruses than the younger ones, an effect not found in the girls. This change is in accord, to an extent, with general brain growth patterns, where gray matter grows until the early teens in males and females (stopping slightly earlier for females than males) and then begins to decrease into adulthood. However, the females' gray matter in the straight gyrus did not decrease with age and the males' did. Finally, and most interestingly, the interviews and assessments of the study subjects (in this and previous similar studies) identified a relationship between the straight gyrus and self-described/interpreted femininity. In adults, higher self-rated association with feminine traits was associated with larger straight gyrus volume. In children the opposite happened, that is, higher self-association with femininity was correlated with smaller straight gyrus volume. Not a particularly clear outcome, aside from the apparent connection between the size of the straight gyrus and self-reported femininity. The researchers conclude that "the origins of the relationship between sexual dimorphism of straight gyrus morphology and social cognition have not yet been elucidated."[20] So, there are some potential differences in the straight gyrus between males and females, but the clearest association is with a gendered trait, not necessarily sex.

For over one hundred years the corpus callosum was supposed to be the Holy Grail of brain differences between males and females (and, earlier in the twentieth century, between human "races"). The corpus callosum is a broad bundle of millions of nerve fibers that lies under the cerebral cortex (the convoluted outer layer of the brain) and runs along its midline. The central part of the corpus callosum is often said to lie on the dividing line of the brain and its nerve fibers reach out like tendrils into the parts of the brain acting as the mediator of signals between the left and right hemispheres. Anne Fausto-Sterling suggests we see the corpus callosum as "a bunch of transatlantic telephone cables. In the mid Atlantic they are bundled. Sometimes the bundles bunch up in ridges, but as cables they splay out to homes and offices in North America and Europe, they lose their distinct form . . . these in turn subdivide, going to separate cities, and ultimately to particular phone connections." She continues, "at its connecting ends, the CC [corpus callosum] loses its structural definition, merging into the architecture of the cerebellum itself."[21] We know the corpus callosum plays an important part in information transfer in the brain, but is it sexually dimorphic?

In the 1990s a number of publications purported to show size dimorphism in the corpus callosum. The general argument was that a larger splenium (the rear part of the corpus callosum, where it is at its thickest) would indicate a better set of connections and maybe reflect better kinds of social or empathetic skills. The argument was that women have a larger splenium than men, and thus better integrative, or holistic, thinking skills. In 1997 the psychologists Katherine Bishop and Douglas Wahlsten examined studies on the corpus callosum and came to the following conclusion: "A meta-analysis of 49 studies published since 1980 reveals no significant sex difference in the size or shape of the splenium of the corpus callosum, whether or not an appropriate adjustment is made for brain size using analysis of covariance or linear regression. . . . The widespread belief that women have a larger splenium than men and consequently think differently is untenable."[22] This seems pretty straightforward: if you conduct a serious overview, the patterned differences disappear. The problem is that so many studies show so many different patterns. A large part of the reason for this is that brain studies generally rely on a low number of subjects (the study with seventy-four subjects is one of the largest!), so as you grow the dataset the actual patterns emerge, whereas with just a few subjects the potential for bias is very large. Looking across all the published studies prior to 2000, Anne Fausto-Sterling found that the majority actually report no sex differences, even when you break them down by specific subareas of the corpus callosum, and for almost every area where at least one report has females with larger structures, another shows no difference.[23] The structure of the corpus callosum makes accurate measurement difficult, especially across studies. The corpus callosum is an important part of our brain, and might hold some cues into human variation in behavior, but at this point in time, Bishop and Whalsten's statement holds. The corpus callosum is not going to tell us about differences between men and women; instead, it tells us that the variation is between individuals, not sexes.

Biology of attachment and attraction: what hormones are at play

It is often asserted that, by nature, men are aggressive and women are nurturing, and that there is "chemistry" between males and females that leads them to desire one another. So one area where we might expect to see biological sex difference would be in the physiological systems of attachment and attraction, especially as they relate to hormones.

We already know that men are not always more aggressive than women, and when they are it is not clear that it is their nature, not their size and culture, that is the best explanation. We know that women do give birth and lactate (and men don't), but we also have noted that humans seem to have evolved a particularly cooperative system for taking care of their young, where males do a lot of caretaking, unlike with most mammals. So, in humans both males and females participate in taking care of their young, but are women biologically more nurturing? We've already established that males and females have all of the same hormones, just that there may be differing levels of those hormones between the sexes. In the context of nurturing we know that a hormone called oxytocin is important in females after they give birth to help facilitate nursing and establish physiological bonds between mothers and infants. We also know that this same chemical is involved in facilitating the physiological bonding between human partners (and in many mammals). This hormone has been moderately well studied in females, and less so in males.

Biologically oxytocin appears to function more or less the same in males and females. However, in females it is also associated with facilitating lactation (milk delivery in response to infant suckling), so this is one difference between males and females in oxytocin function. Oxytocin's overall impact seems to be in mediating and rewarding social attachment via helping induce physiological stress reduction, muscle relaxation, and some neurochemical rewards. This process works the same way in male and female humans.[24] The more secure and positive individuals feel in their relationships the larger the measurable increase in oxytocin during social interactions. While this effect appears in both sexes, a few studies suggests greater health benefits (reduced stress and other cardioprotective benefits) from increased oxytocin levels for females and/or a slightly higher sensitivity to oxytocin in females.[25]

We know that on average males have higher circulating levels of testosterone than females, but there are very few studies that test a group of comparable males and females doing the same things at the same time. The relationship of parenting and testosterone is suggested to be a negative one: interacting with infants can decrease testosterone levels. This appears to be the opposite pattern for another hormone important in caregiving, prolactin. When mothers are lactating and interacting extensively with young infants their levels of the hormone prolactin are high and their levels of testosterone are low. Generally males' prolactin levels are highest in the morning (still much lower than females) and decrease during the day, and we already know from chapter 5 that males' testosterone levels can be affected by activity patterns, dominance interactions, and aggressive events or contests. Recent research looked at the prolactin levels and testosterone levels of males as they interacted with infants. Although there is some variation in the results, the trend was for testosterone to go down and prolactin to stay steady or increase in fathers when they interacted with infants relative to control males who did not interact with infants.[26] Also, testosterone was generally lower in newly married men, married fathers, and men in long-term relationships than in single men.[27] So, while there are differences in the levels and some of the outcomes of these hormones in males and females, social contexts, especially those dealing with attachment and parenting, seem to elicit similar general patterns of hormone response in both sexes.

An important physiological difference between the sexes in many mammals is the presence of pheromones. Pheromones play important roles in attracting mates, sexual behavior, and in inter- and intrasexual conflicts in many mammalian species. There are a number of popular studies that purport to demonstrate the presence of human pheromones (specific chemical odor signals produced by humans). In particular, there has been significant interest in sex pheromones. The best known cases are females detecting specific cues of attractions from male sweat (via the t-shirt experiments). In these experiments females are given t-shirts worn by males for a few days, recently washed shirts and never-worn shirts (or some similar variation). Some of these studies report that females who are ovulating select the t-shirts of either very symmetrical men or men who have immune systems that complement their own.[28]

Few researchers argue that pheromones are involved; instead, they argue that sweat contains indicators of overall health, which is supposed to be related to symmetry or better immune systems. The other well-known case is the reported instance of women living together and having their menstrual cycles synchronize.

The only identified and replicated (by two research teams) human pheromone is androstadienone (a steroid that appears in some sweat), which is reported to produce a positive response from females. A recent study also demonstrated that androstadienone acts in both males and females to enhance the ability to focus on emotional cues.[29] Thus, it is not currently clear what sort of differences between the sexes this compound demonstrates (if any). The second case noted above, female menstrual synchrony, has only been reliably supported by a single study with twenty female participants. Additionally, it appears that women produce certain aliphatic acids in vaginal secretions during fertile parts of their cycles (these have been shown in other primates to attract males), but human males do not seem to respond to them in any consistent manner. The bulk of studies seeking to identify and validate pheromones in humans either refute their existence or give nonsignificant results.[30] This makes it unlikely that there are significant sex differences in pheromone cues of attraction, or at least that, as of the current moment, we do not have any robust evidence of such.

Male and female biology: we have differences, but we're made of the same stuff

Male and female bodies have many differences, but they overlap extensively in structure and function. Looking at average differences blinds us to the important system-wide view, the normative range of variation, and how bodies actually function. When we look at the biology of males and females we are constantly reminded of one major point: we are all *Homo sapiens*. One can easily focus on the clothing, the hairstyles, the cultural behavior, the social history, and the modern-day ideas about gender and being masculine or feminine and see substantial differences between men and women, but very few of those elements match the actual biological patterns in our species. Males are often larger and more muscular than females, and aspects of our skeletons are variations on a theme. This size difference and the slight difference in the way we walk mean a lot to us socially, but biologically these are extremely minor differences.

There is a major difference when it comes to the ways in which our reproductive tracts function. However, we can also see the immense similarities underlying these differences. The tissues that make up the reproductive tract are the same in males and females—it is the same stuff that undergoes development but with different endpoints. The hormones that affect the functioning of the reproductive system are the same for males and females with varying levels and patterns found between and among them depending on social context, age, behavior, and other factors. In a chemical and physiological sense hormones frequently act the same way in males and females. In attachment and bonding the functions are the same with similarities in response patterns, but some differences are found in levels and contexts of hormone action.

One could argue that if there were really deep-seated differences between male and female human behavior and biology they should show up in the brain. The genitals start in the same place and end up looking different, the brain does not. Our brains, rather than being very different, are pretty much the same. Aside from the size difference, maybe some differences in the area of the straight gyrus, and the fact that females' brains stop growing earlier than males (as with every other part of their respective physical bodies), there are no consistent and replicated reliable differences in the male and female brain; it is a human brain.

Looking at the body, reproductive systems, hormones, and the brain, it is obvious that the sexes are alike as much, if not more, than they differ. The myth that males and females are biologically very different from one another is busted. This does not mean that men and women do not differ from one another. I am not arguing that males and females are the same, rather that we are humans and that the actual biological differences between the sexes are much smaller than the behavioral differences between the genders. Understanding this distinction, sex and gender, is core to busting the myth.

MYTH BUSTING: BEHAVIORAL DIFFERENCES BETWEEN MALES AND FEMALES ARE NOT AS GREAT AS WE THINK THEY ARE, NOR ARE MOST DUE TO OUR EVOLUTIONARY HISTORY—CULTURE MATTERS AND GENDER COUNTS

Harkening back to chapter 2, remember how powerful the societal shaping of behavior is. Of course biological differences between the sexes do lead to some adult differences, but having demonstrated that these biological differences between the sexes are smaller or less extreme than we may have thought we can move forward to think about what gender is, what gender differences are, and how they emerge. But first we need to ask the question, how much do males and females actually differ in behavior and skill?

The Gender Similarity Hypothesis

It does appear that on many, many different human attributes—height, weight, propensity for criminality, overall IQ, mathematical ability, scientific ability—there is relatively clear evidence that whatever the difference in means—which can be debated—there is a difference in the standard deviation, and variability of a male and a female population.

—Lawrence Summers (former president of Harvard University)[31]

In the previous section we laid out biological differences and similarities between men and women; now we are interested in behavior. Is Lawrence Summers correct? We know that men, on average, are taller and heavier, but are men and women really different when it comes to IQ or mathematical and scientific ability? Height and weight differences are part of our biology and evolutionary heritage, but can evolutionary differences explain male-female differences in skill and behavior? To answer these questions we need to bust a specific part of this myth first: just exactly how do males and females differ in behavioral potential?

In our society we often think about sex differences in the context of a specific set of skills: verbal, mathematical, spatial perception, and assertiveness. These are all variables that are commonly assessed on psychological and standardized exams. In our society we have very specific assumptions (and expectations) of differences between males and females in these areas that fall in line with Summers's comments. Many anthropologists have long held that male and female behavior varies across cultures, and none would disagree that in general one can point to many differences in the social roles and behavior of males and females. But do these behavioral patterns based on social roles reflect consistent and identifiable differences in the behavioral potential and actual skills of humans? That is, are they tied to biological, evolutionary differences between the sexes? To answer this question, we need ask, what exactly are the differences between men and women?

This myth about male-female differences in behavior and potential owes a large part of its history to the famous meta-analysis published by the psychologists Eleanor Maccoby and Carol Jacklin in 1974; they reviewed more than 2,000 reports of gender differences and found that most societal assumptions about differences were not supported, that males and females were much more similar in behavior and potential than previously thought.[32] However, they did argue for a set of differences in four specific areas: verbal ability, visual-spatial ability, mathematical ability, and aggression. It is this assertion about differences that has pervaded our mindsets for nearly forty years . . . we tend to forget that Maccoby and Jacklin's main point was about gender similarities.

In her recent work, the psychologist Janet Shibley Hyde emphasizes Maccoby and Jacklin's main point: "The gender similarities hypothesis holds that males and females are similar on most, but not all, psychological variables. That is, men and women, as well as boys and girls, are more alike than they are different." [33] She goes on to suggest that we can take a look at the psychological literature, at the bulk of the actual published data from the kinds of tests that specifically focus on male-female differences, to get a good idea of how much men and women actually differ in their abilities. In her 2005 study she took an overview of psychological and standardized assessments of cognitive variables (math, verbal, spatial), communication (verbal and nonverbal), social and personality variables (aggression, negotiation, helping, sexuality, leadership, introversion/extroversion), psychological well-being, motor behaviors (throwing, balance, flexibility, etc.), and a few others (moral reasoning, cheating behavior, etc.).

Shibley Hyde's data consisted of examining 46 previous meta-analyses of male-female differences (published between 1980 and 2004), consisting of nearly 5,000 reports and assessing 128 psychological measures.[34] In comparing the reports Shibley Hyde uses the d measure, which reflects how far apart the male and female means are in standardized units.[35] As with the earlier discussion about men's and women's heights, remember that the closer the means the greater the overlap in the overall ranges. Shibley Hyde argues that the gender similarities hypothesis would be supported if "most psychological gender differences are in the close-to-zero ($d \leq 0.10$) or small ($0.11 < d < 0.35$) range, a few are in the moderate range ($0.36 < d < 0.65$), and very few are large ($d < 0.66 < 1.00$) or very large ($d > 1.00$)."

What did she find across the 5,000 reports in 46 meta-analyses? For 78 percent, the d measures are close to zero or small (38 percent: $d \leq 0.10$; 40 percent: $0.11 < d < 0.35$). Where are the large gender

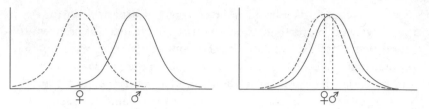

FIGURE 8. Graph on the left shows differences in male and female heights; graph on the right shows male-female psychological gender differences with *d* = .35. Adapted from L. Eliot (2009) and J.S. Hyde (2005). Adapted from L. Eliot (2009), *Pink Brain, Blue Brain* (New York: Houghton Mifflin Harcourt) and J.S. Hyde (2005), The gender similarities hypothesis, *American Psychologist* 60(6): 581–92.

differences? Males scored noticeably higher (*d* > 0.35) in grip strength, sprinting, throwing velocity and throwing distance, masturbation, views on casual sex, physical aggression, and mental rotation of objects. Females scored higher on indirect aggression (reinforcing Archer's studies discussed in chapter 5), agreeableness, and smiling. Thus, the male and female differences in behavior and potential (especially the math and verbal abilities) are aptly demonstrated by comparing height overlap with the pattern of overlap on the vast majority of Shibley Hyde's comparisons (figure 8). Looking at the two graphs, one can see that the vast majority of the assumed male-female differences in the psychological and skill variables overlap extensively. Even more impressive, this graph is the overall mean of the entire dataset, and 77 percent of the actual measures have even less difference and more overlap than shown in the graph!

This is an excellent example of some real biological differences between males and females, which might relate to some behavioral differences (physical aggression or strength-related activities), but the vast majority of psychological tests (and the fact that our brains are the same) demonstrate, unequivocally, that males and females are much more similar than they are different in behavior and ability. A few differences stand out as interesting. Males are generally better at mental rotation of objects, for example. Some have argued that men evolved better spatial skills due to their selection for hunting ability. The problem with this explanation is that males don't do better on all spatial skills, such as mapping skills or spatial memory, but just on certain mental rotation of objects; moreover, in many societies females also hunt (just not large game). It is not clear what that means. But it is absolutely clear that men did not evolve better math skills and women better verbal skills—the data refute that assumption.

But other differences do remain: physical aggression in males versus indirect aggression in females, females' greater smiling and agreeableness. In chapter 5, we suggested that the aggression differences are in part explained by physical size and muscle strength. But smiling and agreeableness? What if these differences do not have their origin exclusively in our evolutionary past, but rather in our cultural present? What if these differences, even those of aggression, are due the ways in which we as humans live in and use culture? I suggest that the anthropological concept of gender can help us understand this possibility.

What is gender?

Gender refers to the social, cultural and psychological constructions
that are imposed on the biological differences of sex.
—Serena Nanda (anthropologist)[36]

. . . the formation of gender roles, by which people of each sex are
expected to have psychological characteristics that equip them for
the tasks that their sex typically performs. . . .
—Wendy Wood and Alice H. Eagly (psychologists)[37]

So we see that males and females are not that different in skills on tests but we do recognize that men and women do differ in our society, in our daily lives, in how we see each other and expect one another to act. Anyone reading this book will be able to describe more or less typical behaviors for males and females in their own society. Why? One word: Gender.

In general most people, and many researchers, use the words "gender" and "sex" interchangeably. The two are related, entangled even, but not the same thing. Anthropologists have long held that gender is best seen as the culturally influenced perception of what the sexes are and the roles they are expected to play. Sex is a biological definition (XX or XY . . . more or less) and gender is how the social worlds, and expectations, of the sexes play out. Gender is best conceived of as a continuum, not a dichotomy.[38] At one extreme end we have total femininity and at the other end total masculinity, with most people falling in between those points. In our society, we expect sex-females to fall largely toward the behaviorally feminine side and sex-males to be mostly toward the masculine side. That is, behaviors we culturally associate with masculinity, like assertiveness, aggression, intense interest in athletics, are seen as being normal for the male sex. So when women exhibit these behaviors we see them as behaving like men on the gender spectrum. The same is true for men who exhibit socially feminine behavior such as heightened displays of emotion, subservience to others, intense interest in Broadway musicals or daytime soap operas; we see them as being like women. These examples are very stereotypical, and there are many, many exceptions to this pattern, but I choose them for a reason: everyone reading this book has a social context of gender that enables them to understand these specific examples.[39] Gender works because it is a core part of the social fabric in which we develop our schemata, the way we see and interpret the world.

The gender roles of society reflect a kind of division with sex-females expected to fill particular roles and sex-males other ones. There can be a good deal of overlap, but to a large part the gender pattern holds. This is true for social roles, in marriage (the male is supposed to ask the female), in public (females can cry at a sad movie, men are supposed to be stoic and comfort the females), and for more formal roles such as paid jobs. As an example of this, think of jobs we consider female (secretaries, librarians, nurses) and those we think of as male (construction workers, business managers, airline pilots). What do you imagine when you picture each one of those jobs? For example, when you picture a pilot most of you will see a male, although there are many female pilots. There are many jobs in which both sexes participate, but there are many gender-based differences when we associate the job with gender. Picture a lawyer, now picture a female lawyer, and now, a male lawyer. In the first and third instances most likely you pictured a man and in the second a woman (though some of you might have pictured a woman in the first case as well). But are they dressed the same? What about hair and accessories? What are they carrying and what kind of shoes do they have on? How would you expect them to behave in the courtroom? The point is we have specific expectations of how the genders should look and act. These expectations are a central part of our culture.[40]

There is also a very strong association between sexuality and gender. We have expectations based on gender roles about how males or females should feel and think about sex. In our society, the roles one plays in sexual activity and the ways in which one displays sexuality in public are highly gendered. In our society (and in many, but not all, societies) homosexuality is often associated with gender transgressions. We tend to socially classify homosexual men as feminized and homosexual women as masculinized, regardless of their actual behavior. This is because we have a particular set of dichotomized expectations associated with gender when it comes to sexuality; men are men and women are women, and they have different relationships with sexuality, often seen as complementary. We expect one partner in a sexual relationship to act "female" and one to act "male" in the kinds of gendered behavior exhibited. Same-sex couples may challenge our expectations because most of us so tightly associate gender with biological sex. In the final myth-busting section of this chapter we examine the actual data for sexual behavior and sexuality to see how they fit the cultural expectations for the sexes and genders.

Serena Nanda, an anthropologist who specializes in gender, points out that while it is easy to think about a sex-gender dichotomy, this creates an artificial nature-culture divide. As we have illustrated in every chapter in this book, humans are amazingly complex. We are naturenurtural, so making a clear distinction between biology and culture is very difficult in many cases. It does not help for us to think of sex as biologically fixed and gender as culturally contingent, like a flexible behavioral cloak thrown over biology. Nanda reminds us that using such a dichotomy ignores the "integration of biology and culture in human life, experience, and behavior." She opts for the term sex/gender to best describe what we are actually talking about. The two are perpetually intertwined, just not always in the ways we think. So it is best to think of sex/gender as a dynamic system of interaction rather than one physical part (biological sex) and one cultural part (gender); you can't have one without the other.

It is gender differences that we are interested in trying to link to our evolutionary past. We know that size and strength differences might explain some of the patterns we see in males and females. But these are small and we know that as individuals there is enormous overlap in behavior and potential between males and females in behavior and biology. Are some of the larger gender differences introduced and maintained in ways that are not related to our evolutionary past? Do our current societies create and maintain some of these differences?

Development and Maintenance of Gender Differences

While we see infants through gendered eyes, the infants themselves do not have full-blown gendered behavior and perceptions at birth; instead, they have to acquire gender as they develop. Given Nanda's point above, we need to think about gender acquisition as part of the biocultural development of the human being. In young infants (by about 1.5 years of age) the gender schemata begin to develop, with gendered play patterns emerging by about two years of age. These patterns differ by culture, but one consistency is related to size and strength. Males start to play in a more rough-and-tumble manner than females at about this age (on average, there is a lot of overlap still). By ages three to four children begin to display consistent gendered behavior and at six to seven years children form relatively fixed gender stereotypes and behave more or less in accord with them. Each child develops his or her gender in the context of a given society, so the specifics of masculinity or femininity will vary for children depending on the societal norms. This is why we see overall similarities in gender within cultures but interesting differences between them. These differences can be in the ways in which the genders interact in public or in mixed-gender contexts, how same-gender individuals act around one another, the ways in which sexual behavior is initiated and carried out, the changes one undergoes socially before and after marriage, and so on. However, each individual human develops with a specific set of biological and social conditions resulting in culture-wide gender similarities but also, very importantly, a wide range of individual differences in gender behavior.

The psychologists Wendy Wood and Alice H. Eagly argue for a biosocial approach to gender that attempts to fuse biological and cultural developments together to better understand certain patterns in gendered difference. They looked at anthropological records of hundreds of societies and examined the gender roles, division of labor, and patterns of sex/gender differences over time. Keeping in mind that there are male and female size and strength differences and that females give birth and lactate, they also looked at the different ways of living (foragers, agriculturalists, pastoralists, industrial societies, etc.) and noted the different divisions of labor inherent to them. They looked at different types of social and kinship systems as well. They found that differences largely came from interactions between the physical specialization of the sexes, like female reproduction, and the economic and social structural aspects of societies. Their biosocial approach sees psychological attributes of women and men as emerging via the evolved characteristics of the sexes, their developmental experiences, and their activity in society.[41] The bottom line is that gender emerges from the combination of our bodies, cultures, and individual experiences. Our bodies are shaped by our evolutionary histories, resulting in some important differences, but so are our brains, resulting in important similarities in behavior and potential. What about cultural and experiential impacts?

Wood and Eagly found that there is variation in the roles males and females play across societies, with high degrees of overlap in many areas, but greater differences being found in aspects of those societies that deal directly with size and strength (such as large-game hunting with spears) or giving birth and taking care of young children, and that other patterns become associated or emerge from, these differences. They suggest that much of the current social division of labor we associate with gender emerges from both the biological facets of being human and human evolutionary histories combined with our histories of resource use and distribution.[42] These assertions are supported by the fossil and archeological record of human evolution and by the fact that gender roles and the division of labor have undergone substantial changes over the last few centuries as societies change both structurally (industrialization and technology changes) and socially (political and educational changes).[43]

But what about today's society? Are there broad-scale social patterns that reflect gender differences (or cause or reinforce them)? As adults we see a wide variety of societal differences between males and females that are not directly tied to interpersonal behavior patterns, but rather to the ways in which societies structure themselves and are governed. These patterns can act to create and maintain differences between males and females, because in each case males tend to have higher access and control over these categories. The differences are in areas such as social and political power, economic power, educational status, and health, as reported in the *Global Gender Gap Report 2010,* which tracks progress over five-year spans: "On average, over 96% of the gap on health outcomes, 93% of the gap on educational attainment, 59% of the gap on economic participation and 18% of the gap on political empowerment have been closed. No country in the world has achieved gender equality."[44]

The fact that aspects of our societies are biased toward male control is not part of our evolutionary heritage, it is part of our cultural reality. There is a gender gap in economic and political power that constructs and helps maintain gender roles and inequality. The World Economic Forum is an agency in Switzerland that collaborates with researchers at the University of California, Berkeley, and Harvard University to assess the division of, and access to, resources between males and females in 134 countries. Controlling resources (political and economic) enables the control of most major aspects of our social lives. What does the gender gap look like in the United States? In 2010, the United States moved to 19th place (up from 31st in 2009), with an overall gap index of 74 percent (Iceland was number 1 with a gap index score of 85 percent and Yemen was number 134 with a score of 46 percent). This overall score for the United States reflects the percentage attainment by females relative to males in the areas of interest.[45] This increase in the rankings is largely due to the fact that in the United States females and males are near parity in educational attainment (we are tied for number 1) and participation in the workforce (we are tied for 6th place). This was not true a century ago. Cultures are dynamic and change rapidly. However, larger differences still remain in earned income and wage inequality (the United States is 64th in this measure) and political empowerment (the United States is 40th here).[46] Men make more money for similar work, hold more positions of power, and predominate in political roles. This is a global pattern, but not an evolutionary one. There are no patterns of biological or behavioral differences between males and females that make males run companies or societies better. These are aspects of societal structures that act to maintain broadly held ideas about gender. When children grow up within a society, they acquire the templates that are around them and these help create their schemata. These contexts set the stage for our biosocial development, resulting in what we experience on a day-to-day basis.

There is no evidence that most gender behavior and the gender gap reflect evolved patterns

Males and females have important biological differences and important gender differences, but they have even more similarities. It makes sense that these similarities are due to our evolutionary history as humans. Both in biology and in behavior and potential the differences are smaller than we generally think they are, and only a few can clearly be linked to aspects of our evolutionary past.

In the areas of gender aggression differences, it seems clear that males' size and strength are important factors in their increased likelihood of exhibiting physical aggression. However, as discussed in chapter 5, the details are quite complicated. Women also use physical aggression, at even higher rates than men, at least within couples. However, males' potential to do greater harm is there. Might this be a reflection of our evolutionary past? Yes. Male size and muscle mass are part of our evolutionary heritage, but this pattern did not evolve so that males could beat up or intimidate females.[47] However, this difference can have an effect in our societies and our gender systems. In social structures where males have political and economic power they can also exploit this physical difference to help maintain these patterns of control. It might be in this case that males' use of physical aggression toward females is a cultural co-option of a biological potential and not a specific evolutionary adaptation in our species. If we think about females' greater use of indirect aggression, the picture is more complicated. Do they use it more because they are on average smaller than males (but then why do females in couples use physical aggression)? Or is this a reflection, like their greater use of smiling and agreeableness, of gendered expectations of behavior? It appears that rather than being clearly evolutionarily linked, many of the actual differences appear to emerge from the structures and expectations of the gender systems in which they occur.

Rather than hanging our hat on a few biological differences and trying to use them to explain gender differences we should be paying attention to what actual gender differences in behavior and potential are and seeing how they relate to our biology and our societies. At the same time we need to realize how much overlap there is across the genders and how variable individuals are in the ways in which they embody and experience gender patterns. Of course some evolutionary patterns have led to gender differences, but very few. The power of cultural traditions, beliefs, and expectations are very strong. Societies do incorporate biological patterns into gender roles, but we cannot look to our evolutionary history to explain the gender gap or most of the general expectations of gender behavior we rely on every day in our society.

Gender behavior is best seen as the result of biosocial development: culture matters, gender counts, and we are simply not as different as we think. Given what we know about male and female behavior and potential the myth that the genders differ dramatically in behavior and potential and that the majority of the behavioral differences that do occur between males and females are evolutionarily hard-wired is busted.

MYTH BUSTING: WE ALL NEED LOVE, BUT NOT NECESSARILY SEX, MARRIAGE, OR MONOGAMY

Romantic love is one of three basic brain circuits that evolved for reproduction: the *sex drive* motivates all of us to look for a range of partners. *Romantic love*, the elation and obsessive thinking that is produced when you first fall in love, focuses our mating energy on just one individual. Following that, *attachment* sets in, the calm and security you can feel with a long term mate, enabling you to sustain your relationship to rear your children as a team. Romantic love is the most powerful, and the beginning of the cascade . . . romantic love is a drive, an instinct that arises from primitive parts of the brain.

—Helen Fisher (anthropologist)[48]

There is a basic story told by many evolution-minded folks interested in human relationships: the body is wired to find mates. Once the best biological mate is found the brain and hormones kick in to create a particular kind of attachment drive: romantic love. This leads to the monogamous pair bond (which may or may not last), offspring, and the natural family unit—a man, a woman, and their children. When you meet the right person for you the evolved chemical cascade will lead you toward a pair bond relationship.[49] The nonevolutionary version (whether religious or secular) is pretty much the same: just remove the chemical part and replace ideal biological mate with spiritual or soul mate. Underlying both of these scenarios is the assumption shared by many evolutionary psychologists as well as the Judeo-Christian-Muslim religions and most people in the United States, that the bonded male-female pair (with offspring) is the evolved, or natural, unit of the human family; that marriage is part of human nature; and that there is a specific pair bond partner out there for everyone. Whether one sees this as the culmination of an evolutionary history or as a spiritual reality, this vision acts to justify the role of marriage and the nuclear family as primary to human nature.

There is no real anthropological, biological, or psychological support for the notion that there is a perfect (or reasonably perfect) match for everyone, or for anyone. This section examines the concept that there are specific biological matches for people and that this mated, romantic pair bond is what humans are evolved to seek. There is substantial evolutionary evidence that humans do seek pair bonds (socially and physiologically), but these bonds do not necessarily involve sex, marriage, exclusivity, or even heterosexuality. We will also see that marriage is not equal to evolutionary or physiological pair bonds, that the nuclear family is not the basic unit of human social organization, and that social expectations for the quality and structure of life after attaining these two things can lead to an array of social and psychological problems for people.[50]

There are three parts to this section of the myth: first, that the unit consisting of bonded male-female + kids is the basic unit of humanity; second, that humans are naturally monogamous and that marriage is a reflection of evolutionary origins; and third, that individuals are attracted to a single, specific mate (pair bond mate or soul mate), with whom they are evolved to have sex, marriage, and exclusivity. These assumptions are not really accurate.

What is love?

We often think of the concept of love at the center of understanding romantic relationships. Humans form pair bonds and are frequently in monogamous sexual and social relationships, but that does not mean what you think it does: romantic relationship ≠ love ≠ monogamy ≠ pair bond.

So what is love? Unfortunately, the answer that most people seek involves a philosophical question beyond the scope of this book. However, we can slightly rephrase the question to ask, what is going on in the body when people feel strongly toward one another and why are these feelings so powerful? The initial answer goes back to the section earlier in this chapter on the biology of attachment. We already know that a suite of hormones and neurotransmitters (including oxytocin, vasopressin, prolactin, testosterone, dopamine, etc.) are involved in developing and maintaining physiological bonds between mothers and infants and fathers and infants. This system also functions in the same way between adults. Physical touch, spending intense social time in contact or near one another, and positive social interactions can trigger it. There is an evolved system in humans that uses social and physical interactions, hormones, and the brain to prime the body to feel closer and more attached to another individual. The anthropologist Walter Goldschmidt called this affect hunger.[51] He argues that the basic system that acts to bond mammalian mothers to their infants has been expanded and co-opted in the human species to act as a social and physiological bonding system between individuals of all ages and sexes. This drive of affect hunger enables humans to form and experience types of social bonds that are not found (to the same extent) in other animals, even in other primates. He also argues that it is these bonds which have enabled humans to do better than almost any other organism on the planet.[52]

So the answer to "what is love" in this context is that it is the biology underlying affect hunger, the ability to form multiple, strong social bonds, and part of the human adaptive niche—the evolutionary history that has made us so successful as a species. However, this notion of love covers what is called love between parents and offspring, between siblings or other family members, and between good friends as well as between romantic pairs. Most people when they ask about love are actually only interested in one version of this: the romantic pair. Culturally we see romantic love as separate from familial or friendship love. Unfortunately, aside from a slightly different pattern of some specific hormones brought about by sexual behavior, there is nothing biologically different about romantic love than any other kind of love.[53] The myth that romantic love is essentially (biologically) different from other types of strong attachment is created and maintained by cultural beliefs and our schemata, not our biology. So asking about the naturalness of romantic love misses the boat. What most people are really getting at when they try to figure out romantic love is to explain the specific strong relationship between two people that we call the pair bond.

What is a pair bond?

In the basic biological literature a pair bond is simply a special, predictable relationship between two adults. When researchers look to humans (and some other mammals, especially primates) this is refined to focus on special and predictable relationships between a male and a female that involve tight social connections and a sexual relationship, and usually includes mating and the raising of young. It is often asserted that this pair bond is the basis of human society and that we can look to our evolutionary heritage to see that it is a major, early event in human evolutionary history.[54] However, pair bonds are not exactly what many think they are and they are not necessarily linked to procreation and the nuclear family in human evolutionary history.

More primate species are said to have pair bonds, and monogamous relationships, than any other group of mammals. This assertion is used to argue that humans are expanding on this primate trend and solidifying the pair bond, and monogamy, as our basic social unit. In 1999 and 2002 I published overviews of the relevant datasets on primates and humans, asking several questions: do primates have more pair bonds or more monogamy than other mammals, are pair bonds the same thing across primates, and are pair bonds the same as monogamy.[55] What did I find? Primates (including humans) are not more monogamous than other mammals (in fact it is really rare, as only about 3 percent of all mammalian species are monogamous); there are a number of primate species that live in small groups consisting of male-female plus offspring, some with and some without pair bonds; pair bonds come in a number of different types across primates; and pair bonds are not the same as monogamy. Other researchers have looked extensively at the biology of pair bonding in voles (a kind of rodent) and a few monkey species as well as humans.[56]

From this work it is clear that there are two types of pair bonding that are of interest here: the social pair bond and the sexual pair bond. The social pair bond is akin to what we described above with affect hunger, and can be defined as a strong biological and psychological relationship between two individuals that is measurably different in physiological and emotional terms from general friendships or other acquaintance relationships. The sexual pair bond is a pair bond that has a sexual attraction component such that the members of the sexual pair bond prefer to mate with one another over other mating options. In humans and other mammals pair bonds are developed via social interactions combined with the biological activity of neurotransmitters and hormones like oxytocin, vasopressin, dopamine, corticosterone, and others.[57] In voles and a few other mammals where the biology of pair bonds has been studied, social and sexual pair bonds are frequently coexistent, but in humans this is not the case. Humans have both social and sexual pair bonds, and the two are not necessarily connected.

Humans have extensive social pair bonding across genders and age categories, probably more than any other species. We can have pair bonds with our relatives and our closest friends, they can be with same-sex individuals or different-sex individuals, same age or different age.[58] Humans are also unique in having sexual pair bonds both heterosexually and homosexually. Our sexual pair bonding, like our sexual activity, is not limited to reproduction.

Recent work in the evolution of human social systems has noted the important role of the pair bond. The primatologist Bernard Chapais has mounted a broad comparative approach looking at primate behavior and models of human evolution and argues that the sexual pair bond precedes the nuclear family structure in human evolution, but that its appearance marked a core turning point in the evolution of the human social system. He argues that the sexual pair bond sets the stage for the kinds of parental cooperation that we see today as a core factor in humanity.[59] I, and many others, have also recently argued for a broader core role for social pair bonds in human evolution. Taking a page from the mounting evidence for a key role of cooperation in human evolution, we argue that social pair bonds are a logical and effective way to enhance and expand the social networks and cooperative possibilities in human ancestors.[60] Pair bonds, both social and sexual, in humans are part of complex social networks that emerged as a core pattern in human evolution. Pair bonds can involve sexual relationships (and in a cultural sense, romantic attachments) and are probably involved in what most people experience when they think of romantic love. But pair bonds are not the same as marriage and they are not necessarily connected to monogamy.

So if love per se and pair bonds do not give us clear answers as to human romantic relationships, is it monogamy that is most important when we are trying to figure out romance and marriage? The answer in the biological sense is no but in a cultural sense, possibly. So what actually is monogamy and are humans monogamous?

We are not naturally monogamous, but we are frequently monogamous

Clearly, the notion that women are designed solely for lifelong pair bonding, and that any deviation from long-term monogamy represents a maladaptive response of our pair-bonding system, is at odds with the prevailing evidence that multiple mating is a relatively common—and in some ways preferred—sexual strategy.

—David P. Schmitt (psychologist)[61]

. . . when it comes to monogamy as mating exclusivity, what we see is not necessarily what we get.

—David P. Barash (psychologist/zoologist) and Judith E. Lipton (psychiatrist)[62]

Over the last three decades sufficient overviews of human mating patterns and sexual behavior have emerged to resoundingly demonstrate that most humans, today and in our evolutionary past, did not mate monogamously across their life span. But many individuals do have one or more relatively monogamous sexual pair bonds during their lifetimes. The majority of cultures today legally sanction both polygynous (multiple wives) and monogamous marriage systems. There is also a robust body of evidence that monogamous marriage systems are not the same as monogamous mating systems, that is, even within monogamous marriage systems there is a good deal of polygamous (multiple partners) mating going on.[63]

In a biological sense monogamy is defined as exclusive mating between two adults across one reproductive cycle. Often the definition also includes the production of young by the two adults. Long-term monogamy would then be exclusive mating across multiple breeding seasons. For nonseasonally breeding animals and humans, monogamy means exclusive mating between two adults resulting in one or more offspring. This is the biological definition, and this type of mating system is extremely rare in the animal kingdom. Even when a species is monogamous socially and in most matings, it appears that between 10 to 20 percent of all matings and a similar number of offspring are the result of extra-pair copulations.[64]

In a cultural sense monogamy is usually assumed to be an exclusive association between two adults, sanctioned by marriage. Often extra-pair sexual encounters by individuals in this arrangement are punishable by civil or religious law. Right away we can see that there is a particularly glaring problem between the biological reality of mating patterns and the cultural assumptions (and laws) of marriage patterns (under which mating is sanctioned).

There is an extensive body of research looking into the history and structure of marriage systems throughout the world, too voluminous to review here. Basically, anthropologists, historians, and sociologists agree that in general marriage (in both secular and religious systems) is best seen as a social system for legitimizing reproduction and inheritance of property, control of and regulation of sexual activity, and, recently, the culturally sanctioned outcome of romantic love.[65] This is also an important way in which cultures can officially recognize and sanction the sexual pair bonds that characterize human beings.

It is critical to mention that the current view of marriage that dominates the Judeo-Christian-Muslim religions, and the cultures that are intertwined with them, is a fairly recent occurrence in human history.[66] This is the idea that romantic love and marriage are connected and that marriage is the ultimate outcome for a couple in love. It begins to emerge in the sixteenth century and rapidly spreads across much of the Western world, and now much of the globe.[67] Previously, and in many societies still today, there is no necessary connection between romantic love (or lust) and marriage. Today, most people in the United States do identify marriage as a natural goal for humans, and at least in public, equate monogamy and marriage.

The bottom line is that there is a difference between marriage and mating (or at least sexual activity). True or long-term monogamy is rare in an evolutionary sense and not the typical mating pattern for humans. However, monogamy, via the proxy of marriage, is the expected cultural norm in many societies. And, importantly, most humans today who are married are in assumedly monogamous marriages. At the same time humans do socially and sexually pair bond, but are all married couples sexually pair bonded? And/or socially pair bonded? Given the enormous variation in why and how people marry, probably not. But there is very, very little research asking these questions. We currently have no data on this critical measure.

However, we do know that there is discordance between the biological patterns of sexual relations and attachment and a society's cultural expectations. For example, if a married pair is not sexually pair bonded with one another, it would not be surprising that they would have trouble being sexually monogamous. If a pair is socially pair bonded, it might not matter to them that there are occasional (or frequent) deviations from sexual monogamy. However, even if a married couple is sexually and socially pair bonded the basic biology of human mating predicts that over their time together each will have occasional physiological and psychological desire and/or inclination to mate with other individuals. Our cultural expectation of sexual monogamy is at odds with our evolutionary heritage and basic biology. However, our expectation of social monogamy is generally reflected in the biology and behavior of social pair bonds between partners.[68] The real wrench in the system is sex, not social relationships, and this has to do with sexual behavior, desire, and attraction.

It is not human nature to seek marriage and a specific sexually monogamous romantic relationship, but it is in our nature to pair bond and in our culture to seek marriage

Human affect, hunger, and the need to form multiple physiological and psychological close bonds with other humans is core to who we are. It is part of our human nature. If Walter Goldschmit is right, and this is what we call love, then the human need for love via social pair bonds is a hallmark of our evolutionary history and current biology. Humans are rarely sexually monogamous over their lifetimes. Rather we can form multiple sexual pair bonds of differing durations over the course of our lives, which may or may not also be social pair bonds. In the next section we also point out that humans can have sex (and do) without pair bonds at all.

There is no good evidence that pair bonds evolved because of the nuclear family (or for the initiation of one). In fact, there is evidence that pair bonds preceded the more recent pattern of one male and one female plus their offspring being a central residential and familial unit in our species. All of this conflicts with the widespread cultural expectations (built into our schemata) that people can all hope to find a monogamous partner, be monogamous sexually, and that once we enter into the nuclear family relationship we are set. The myth that it is a natural human goal to obtain a unique and powerful specific sexually monogamous romantic relationship and a nuclear family is busted.

MYTH BUSTING: MEN, WOMEN, AND SEX—THE GOOD, THE BAD, AND THE COMPLICATED

Sexuality is a somatic fact created by cultural effect.
—Anne Fausto-Sterling (biologist)[69]

Human sexuality is one of the best examples of our naturenurtural reality. We've already talked about male and female biological and gender similarities and differences, about the biology of attachment, about pair bonds, and about the intricate connections between cultural context, experience, and gender. However, in attempting to talk about sexual behavior we confront a situation where any individual's sexuality is such a biologically and culturally entangled reality that describing in a general way what men and women are sexually is extremely difficult.[70] But we can try.

The anthropologists Hastings Donnan and Fiona Magowan recently completed an overview of the ethnographic and theoretical literature looking at how sexuality and sexual behavior are narrated and embodied (described, lived, and felt) across cultures. Their conclusion is telling:

> . . . it is not possible to pin down sex to any one thing . . . sex can be many things to many people, including but not limited to a blend of personalities, social rules, desire, intimacy and performance, moral order and national

image that speak to processes of sexual embodiment, varieties of sexual practice, and the dynamics of culture.[71]

So it's complicated, really complicated. And the myth of male and female differences in sexual behavior is a dominant one. Because we are primarily interested in debunking this myth (or at least showing how it is really not that simple) we'll only tackle two parts of the picture: a brief review of what we know about how people are attracted to one another and an overview of what males and females actually do in regard to sexual behavior and activity. By looking at what people actually do we can tackle the misinformation in popular perceptions about male and female sexual actions and desires. The bottom line is the same as in every other section: we are not as different as you think, which is a fact well-summarized in the conclusions of the 2010 National Survey of Sexual Health and Behavior: "Men and women engage in a diverse range of solo and partnered sexual behaviors throughout the life course."[72] How males and females do so, the similarities and differences, are what we are interested in. But first, what is it that draws individuals together to have sex in the first place?

How People Are Sexually Attracted to One Another

In order to understand the ideas about how individuals select mates or are attracted to others enough to engage in sexual activity, we need a brief review of the development of sexuality. There is reasonably good evidence that a person's sexual orientation is relatively fixed early in life but that the specifics of any one person's sexuality varies across the life span. In a very general sense the average sequence looks something like this for both males and females.[73] From birth until about three years of age the nervous and endocrine systems are developing patterns and connections, which means that physical stimulation often results in a response in the genitals. There is an early association of physical contact with positive neurochemical feedback and reaction in the genitals. This is often considered nonsexual in the sense that adult sexuality and attraction are not at play for the infant or toddler, but it is setting the biological pathways for sexual response. The initial phase of strong attachment and the beginning of pair bonding (at least with caretakers) starts here as does the initiation of gender identity. Between three years of age and the onset of puberty (usually between nine and thirteen years of age) children engage in sex play within their peer group, both homosexually and heterosexually. This activity is considered largely nonformative in that there is no necessary connection between adult sexual orientation and the exploratory sexual activity of children. This is also the period where intentional exploratory masturbation begins.

From adolescence through young adulthood (mid-twenties or so) there is a variety of activities which influence the full-blown adult sexuality and the completion of the biological and physiological changes of puberty and menarche.[74] Obviously there is enormous variation between individuals but long-term overviews show that on average the following elements emerge in this period: increased masturbation, increased sexual interactions, increased physiological attractions, first coital experiences, and first pregnancies (in females). During this period the following cultural elements are shown to influence the shape adult sexuality takes: gender and cultural role changes, parental impact, peer impact, national and ethnic cultural impact, religious beliefs, and economic, political and other external limitations. Finally, in the adult years (late twenties until death) there are both physiological factors (menopause and related physiological changes in females and decline in function of the reproductive organs in men) and social factors (cultural expectations and restrictions) that influence sexuality and sexual behavior. This overall pattern is more or less the same for males and females; however, females tend to be slightly ahead of males (hit the phases earlier) in this developmental pattern, just as they are on brain and body growth.

As the body and mind codevelop one's sexuality, individuals begin to have patterns of attraction. This means that certain cues or assemblages of traits elicit strong attraction and initiate sexual response physiology. These patterns of attraction are the things that turn people on. This is one area where there might be some important male-female differences.

Some gender differences in attraction are largely shaped by cultural and experiential context: clothing, hairstyles, certain mannerisms or behavior, certain smells and types of language use. Given humans' tendency to belong to a group and participate in that group's social patterns, popular culture and one's peers have a great impact on the development of attraction. While there is extensive individual variation most noticeable gender differences are closely linked to cultural expectations and patterns. There is also often a connection between situation and sexual activity that may not relate directly to general patterns of attraction or mate choice. Use of alcohol, peer pressure, and a variety of other factors can affect actually engaging in sex, but in this section we are primarily interested in male and female patterns of attraction.

Some researchers believe that there are evolved systems of attraction that are the result of adaptation by males and females to focus in on traits that indicate higher mate quality. The argument is that over time those who have the best intrinsic abilities to identify and be attracted to higher-quality mates will benefit in evolutionary terms (more or better-quality offspring).

In their overview of evolutionary approaches to human physical attractiveness, the evolutionary psychologists Steven W. Gangestad and Glenn J. Scheyd suggest that there are certain aspects of attraction that are best seen as the result of human biological evolution.[75] Interestingly, the first of these is a similarity between males and females: unlike many animals where only one sex does the majority of mate choice, in humans there is mate choice by both sexes. However, some results suggest that while both males and females differentiate the desirability of potential mates, they might do so using different cues. Some suggested areas where this occurs are facial features and symmetry, body symmetry, body shape, and immune system complementarity.[76] In some studies men report preferring females with what are considered more feminine faces (small chins, large eyes, high cheekbones, full lips). However, while a few studies report women preferring masculine faces (broader faces, more robust skulls), other studies show no preference at all. Interestingly, a few studies demonstrate that both males and females tend to find digitally averaged faces most attractive. There are also some data that suggest that both males and females prefer more symmetrical faces over less symmetrical ones.

Much research has gone into the assessment of female body form by males, especially waist-to-hip ratio (WHR). It is argued that a slightly lower than average WHR is especially attractive to males and that WHR might be related to female fecundity (high fertility). However, this remains a contested proposal, both in the sense of what WHR reflects and in the sense of cross-cultural contexts, modern media impact, and body shape variation.[77] There is some preliminary evidence that women are particularly attracted to males who have more masculine bodies, but these data are almost exclusively from North American college student samples, so it is not clear if they could reflect a human-wide pattern.[78]

Across both the body and face attraction studies we find a few differences between males and females, but not much. It is very interesting that the differences we do find seem to suggest that the biological differences in size and shape are what are being focused on, and that men might be more influenced by visual cues than women. There also might be some attraction by women to men who have complementary immune systems (but not vice versa), maybe indicating that females are using olfactory cues in attraction more than males.[79]

What is typical sexual behavior?

It is difficult to have ubiquitous conversations about sexuality and sex for pleasure in the absence of accurate data about the actual sexual experiences that are common.

—M. Joycelyn Elders (former US surgeon general)[80]

It is very difficult to study sexual behavior. Even in our closest evolutionary relatives, the apes, it is only recently that we have come to realize that sexual behavior is common, not always related to reproduction, and complicated. For humans, we often assume that men have more sex than women and that men are more interested in sex than women. Is this actually true? In the late 1940s and 1950s the zoologist Alfred Kinsey revolutionized the study of human sexuality by recoding testimonials and interviewing over 5,000 males and nearly 8,000 females. The publications from this dataset rocked the academic and public worlds; people had a lot of sex, they had it in a variety of different ways, and most importantly, males and females both had complex sexual lives.[81] Since then there have been a few broad-scale studies of sexual patterns in the United States and other societies. Let's review the data from the most recent study of the United States to help answer the question of what males and females actually do.[82]

The data come from the 2010 National Survey of Sexual Health and Behavior, a nationally representative study of 5,865 adolescents and adults (2,936 men and 2,929 women ages fourteen to ninety-four) carried out in 2009 by a team based at the University of Indiana's Center for Sexual Health Promotion.[83] The results are as follows:

Masturbation: 55 percent of men reported masturbation in the past month, and 71 percent in the last year; 31 percent of women reported masturbation in the last month and 54 percent in the last year, except those over seventy.

Vaginal intercourse: 85 percent of men in their twenties and thirties reported having vaginal intercourse in the last year, compared to 74 percent in their forties, 58 percent in their fifties,

54 percent in their sixties, and 43 percent in their seventies. For women, 81 percent in their twenties and thirties reported having vaginal intercourse in the last year, compared to 70 percent in their forties, 51 percent in their fifties, 42 percent in their sixties, and 22 percent in their seventies.

Partnered noncoital behavior: Men and women of all age groups reported engaging in oral sex and masturbation with a partner. For both oral sex and partnered masturbation the pattern is practically identical in both sexes: the highest rate of oral sex is in the eighteen to forty-nine age group with a decrease in older age groups.

Anal intercourse: More than 20 percent of men between ages twenty-five to twenty-nine reported anal sex in the last year, with younger and older men reporting much lower numbers. More than 40 percent of men eighteen to fifty-nine years old reported participating in anal sex during their lifetime. For women the numbers are almost identical except that females reported slightly higher frequencies of anal sex over a larger age range (eighteen to sixty-nine) than males.

Same-sex sexual behavior: While not common, this behavior is by no means rare. Across all age categories about 8 to 10 percent of men reported engaging in same-sex sexual activity during their lifetime, with higher numbers (13 to 15 percent) reported in the forty to fifty-nine age groups. About 5 to 9 percent of women report participation in same-sex sexual behavior during their lifetime, with much higher figures (up to 17 percent) for the twenty to thirty-nine age group. One key difference between males and females is that a higher percentage of males reported same-sex encounters in the past month or year than did females (except for females aged twenty to twenty-four and thirty to thirty-nine). This part of the survey did not ask about sexual orientation so it is not clear what percentage of these numbers reflect homosexually oriented individuals as opposed to heterosexual or bisexual persons engaging in same-sex sexual behavior.[84]

One pattern of similarity and difference that emerges in this study is the decline in sexual activity with age (especially over sixty). However, in this decline there is a slight difference between males and females; the female decline is larger than the males'. This is a pattern observed in other studies: as females age their overall participation in sex goes down (on average). This is especially acute in married couples where females' participation in sexual activity with their partners is negatively correlated with the length of time married.[85]

The data from this study show few major differences between males and females in sexual activity. However, one might argue that the real differences between males and females are not in sexual activity but in the expression of interest in the pattern of sexual behavior as it relates to mating. This concept is called sociosexual orientation and is measured via the Sociosexual Orientation Inventory (SOI), which is a self-reported measure of individual differences in human mating strategies. These scores range from low (preferring monogamy) to high (preferring a promiscuous mating). The basic argument is that there is a pervasive pattern of differences between males and females in attitudes about sex, fantasy, and sexual behavior. The assumption is that men should rate higher or more unrestricted on sociosexuality than women because of their evolutionarily based tendency to want to reproduce as much as possible and females' tendency to look for the best mates rather than mate with many males; in short, males want to have many mates and short-term mating investment, and females want longer-term mates and long-term mating investment.[86]

In general the major datasets reporting on this variable show that men across the globe tend to score higher than women on the SOI. In studies of the United States, men do tend to report higher interest in sexual activity and sexual fantasies, higher numbers of preferred or actual sexual partners, and wanting short-term versus long-term mating opportunities (on average). However, are these results best attributable to evolutionary differences between males and females? Where are those differences located? Not in the brain or the body . . . but maybe in the perception of sex and mating/marriage patterns. What influences this perception? Our cultural schemata do. We are naturenurtural creatures and the context in which we develop is going to have enormous influence on something like our self-reported perceptions of sex and sexuality.[87] Even the evolutionary psychologist David P. Schmitt, author of the most comprehensive sociosexuality survey, concludes:

> In the present study, the most consistent finding was that men scored higher on sociosexuality than women across cultures. Several different theories were evaluated concerning why men and women differ in this way. They all received at least some empirical support. As a result, we are left with the relatively unsatisfying conclusion that sociosexual sex differences are predictable from several theoretical perspectives, none of which is conspicuously superior to the others. . . . At present, it appears that multiple perspectives

are required to more fully explain the cultural and gender-linked variance in sociosexuality.[88]

Yes, there is a difference in self-reported perspectives on sexuality, but are those differences as great as many make them out to be? Psychologists David Buss and David Schmitt argued for a radical difference in male and female mating strategies based on self-reported ideal partner number over time. Males reported wanting an average of about ten partners over their lifetime and females reported wanting about four. However, if you look closely at the data and ask what the median was (the absolute true middle of the distribution of responses) the answer came back as around one for both males and females! No real difference. In fact, the large average differences seem to be brought about mostly by more males reporting much higher numbers (a hundred partners or more) than females; these outliers increased the average.[89] Also, all of these data come from US college students, not really a great representative sample of humanity.[90]

So what do these data tell us? First, people have a lot of sex and second, there are relatively few differences between males and females in the kinds and patterns of sexual activity. This result should not be too surprising as 90 percent of the time (more or less) it is males and females who are having sex with one another, so the numbers should be very similar. In the majority of studies presented here, as in Kinsey's study sixty years ago and in the few similar studies that have been produced in the interim since then, the results are generally the same: men and women engage in sexual activity in more or less the same manner. Men might talk about it more freely or express more active interest to questioners than females, but this also might reflect the power of culture and gender. It also appears that females have less sex than males as they age and that there might be a difference in sexual interest over time in married males and females.

Do males just want sex and females just want good males? No, at least not based on studies of sexual activity or attraction. However, cultural and gender contexts might make many kinds of differences appear. There are behavioral differences between males and females in how they act on and think about sex, but there does not appear to be any consistent evidence, aside from self-reports and interviews of sociosexuality, that suggest this is a property of human evolutionary histories. Males and females are just not that different when it comes to sex. But there does appear to be a core role for cultural and gender-based structures in affecting how we see, live, and think about sex and sexuality. These are patterns of social, historical, political, and economic variation, not distinct biological, or "natural," differences between males and females in regard to sexuality. We can see this kind of effect even across the United States, which is relatively free in regard to constraints on gender and sexual activity in contrast to many societies. It stands to reason that there would be even larger differences between the genders in societies with extreme curtailment of female public movement, freedom of expression, or sexuality.[91] As the psychologist David Schmitt sums it up,

> Women never precisely match the sociosexual psychology of men, but women's overall level of sociosexuality comes closer to men's when it is given the chance. The current findings support the view that women's sexuality is often constrained by cultural values and social institutions, and the "true" nature of women's sexuality includes short-term mating desires and some degree of sexual promiscuity.[92]

Given what we actually know about human attraction, human sexual activity, and sexuality itself, we can safely state that the myth that men and women are really different when it comes to sexuality is busted.

MEN ARE NOT FROM MARS, WOMEN ARE NOT FROM VENUS: BOTH ARE FROM EARTH AND BELONG TO THE SAME SPECIES

There are important differences between the sexes: women give birth and lactate, men are usually larger and more muscular, and the levels and patterns of some hormones vary between the sexes. There are also important similarities: our differentiated reproductive organs come from the same embryonic tissues, our bodies are made of the same stuff and structures, our hormones and brains are the same, we are the same species. This chapter intentionally focused on the similarities to illustrate a main point: men and women are not as different as most people think in our bodies, minds, and behavior.

We know that males are not intrinsically better at math and females better at verbal skills, and that the vast majority of the core differences are tied to the basic facts of physical differences in the bodies of the sexes. For the majority of behavioral traits males and females overlap extensively, almost to the point of insignificant differences at the level of sex, but significant differences at the level of individuals. Even for physical traits we know that there is a greater range of overlap than popularly conceived. Both males and females can care for young (and their bodies respond to this in more or less the same way). Humans, regardless of sex, seek to form social and sexual pair bonds.

Humans are very sexual relative to other mammals.[93] We have lots of sex in a wide range of contexts and formats. We know less about desire within and between the sexes, but we do know that the ways in which people have sex and think about sex are extremely connected to the culture in which they live.

These strong similarities in male and female bodies and behavior do not mean that gender differences are not very real and very important. Just like the concept of socially constructed races, the perception and expectation of gender differences are part of all cultures and impact individuals and society. We all experience these patterns of gender difference—and they can fool us into thinking that men and women are so very different by nature.

Different cultures do it in different ways, but certain patterns are relatively consistent. Males tend to control economic and political resources, not because they are evolved to do so or that women are less capable of doing so, but because of the social and historical paths that have favored patriarchy. Women are associated with the domestic sphere and children due to their giving birth and lactating, not due to any inability of males generally to care for offspring. There is no biological mandate that only females care for young and only males care for economics and politics. In fact, it is highly likely that it is the cooperation between parents and other people in the raising of young that enabled humans to be as successful as we are today.

If this is all true (and it is), then why do so many people (researchers and the public alike) make such strong claims about the nature of human sexual difference? For two reasons: first, they focus only on the differences, ignoring the similarities; and second, they forget, or do not realize, that they are seeing everything around them through their own schemata (their all-encompassing worldviews).

If you ignore the massive set of gender and biological similarities or better, put overlaps between men and women and just look to the gender and biological differences, then of course you are going to assume that we must be different by design (either evolution's or a deity's design). However, if you look both to the similarities and the differences you are struck by how complicated the whole picture is and how the differences fall into specific patterns associated with body structure and cultural expectations of gender. You then have to attempt to explain both the differences and similarities, which means you are stuck dealing with the very complicated biocultural nature of humanity. What you do not have handy is a clear suite of evolved differences in behavior between the sexes.

We are all products of our own societies. We are who we meet and grow up with. If we are told from day one that little boys like trucks and little girls like dolls, that women are emotional and nurturing, and that men are assertive and controlling, we will grow up seeing those behaviors around us. Researchers, especially those looking for evolutionary origins of why we do what we do, have to be extremely careful that they do not overlook the structures of modern human societies and our schemata in their quest to understand the big picture. They must be careful not to already know how the world looks and simply seek an explanation as to why the world is this way without first asking the most basic scientific question: is the world really this way?

The myth of anisogamy, that there is a massive and insurmountable burden placed on females due to their reproductive system, and that this burden is simply not the same for males, is strong and leads many to assume that male and female natures are different because of it. This myth is strong because it does seem to fit for many forms of life on this planet, like insects and even some mammals. However, in many organisms, especially in humans, a system has evolved that requires intensive cooperation between males and females—the actual act of gestating, birthing, and lactating is only a small part of the overall reproductive and parenting effort shared by both sexes. Relying on simplistic notions about females' limitations and males' drive to inseminate as many females as possible as a starting point for evolutionary theorizing is simply not tenable.[94] The human situation has a different basis and thus the hypotheses and research questions need to be expanded and resituated.

MOVING BEYOND THE MYTH

So what now? How does this information, the busting of the myth of extreme differences between men and women, impact our daily lives? First, we need to listen to Anne Fausto-Sterling and discard dualisms. Thinking of males and females as opposites is incorrect biologically and socially, so it will not get us good answers to questions. Looking only at culture and social histories or only at biology and evolutionary patterns is also a false dichotomy and will hamper our abilities to ask and answer important questions. We need to be especially careful when using aspects of gendered behavior as reflections of human nature and we need to be aware of our biases, and the biases in our datasets, at all times. As the sociomedical scientist Rebecca Jordan-Young says, "we are not blank slates, but we are also not pink and blue notepads."[95] Our brains are not made "male" or "female" but develop via interactions between the external world and our own sensory apparatus, our bodily systems have important differences but are more similar than they are different, and gendered behavior and gender relations change over time as our social and structural contexts shift and our schemata change accordingly.

If we discard the myth that men and women are so different then we can see the range of individuals more clearly. If we accept that there are many ways to be male and female and that many of these ways overlap, we can be more accepting of a wider range of masculinity and femininity within and between individuals. A nine-year-old male who picks up a baseball for the first time and throws it ineptly is not "throwing like a girl" as his teammates might say. He is throwing the ball like a human who has not been trained to throw a small round ball with accuracy and speed.[96] When a nine-year-old girl plays baseball well, sliding hard, getting dirty, and running out every time she is at bat she is called a tomboy or is described with masculine adjectives. She is being a good athlete, not being like a boy. These are simplistic examples, but the idea has significant impact across all aspects of our lives. Taking this perspective can help reduce conflict for individuals, and their families and friends, who feel that they fall outside of the social expectations for their gender. It can also create a more level playing field when we look at the abilities and behavior of others, not thinking they will perform one way or another because of assumed limitations of their sex. Again this does not mean people do not vary in occasionally predictable ways. However, if we broaden our categories we might just be pleasantly surprised.

Another way the ideas in this chapter might help is with the expectations for love and romance and marriage that permeate our society. There is no evidence that there is a specific chemical/biological and social match for each individual on the planet. There is also no guarantee that any individual will successfully initiate and maintain one or more strong pair bonds socially and sexually across the life span, although many of us probably do desire such relationships. Marriage is not necessarily one of those pair bonds. It might coexist with one, but getting married and having children does not automatically initiate a pair bond. People need to realize this because spending enormous amounts of time and effort with one other individual is very difficult, and if there is not a pair bond it is probably even more difficult. Romance and marriage are not evolutionary adaptations, they are part of our cultural expectations and patterns, which change over time.

Pair bonds are not necessarily lifelong (in fact most are not) nor are they always the same across the duration of the relationship. Humans can have many pair bonds across their lifetimes and, frequently, multiple ones at the same time. Social and sexual pair bonds can be very similar biologically but their social and cultural impacts can be quite different. Being in a pair bond (social and/or sexual) does not mean that either individual ceases to be sexually attracted to (or active with) other individuals. Monogamy in humans is a social contract, not a biological reality. We can be monogamous, but our bodies and minds are not specifically designed for it.[9]

Men and women do not really want different things from life; in the end we are all humans. However, some biological patterns combined with specific cultural contexts can create different desires, expectations, and patterns of behavior. We must realize that each individual may or may not match the appropriate ideas society has for sex/gender but that such variation is normal for humanity. Understanding how we are similar and different and the range of human variation gives us a broader notion of what is natural for humans. There is no evolved battle of the sexes in humans, nor are gender differences and similarities unimportant, but understanding both how we do and do not vary can help us move forward toward a better society.

7

Beyond the Myths: Now What?

This book explores some of the complicated patterns of becoming and being human, with a focus on the issues of race, aggression, and sex. Hopefully, after reading it you are less likely to passively accept popular notions about what it means to be human. Simply asking if there is a human nature is the wrong question. Rather, we need to ask what do humans actually do? How do we vary and how are we same? And, most important, how do we best explain the results of these questions? We have to be ready for multiple valid explanations and intertwined and complicated answers. A one-size-fits-all approach is insufficient but it sure helps propagate myths. These myths limit our abilities to ask good questions about who we are and why we do what we do. There are many commonalities, and important differences, across our species. The objective of this book has been to show that these similarities and differences are frequently not the same as the ones portrayed in popular myths, expectations, and perceptions of human behavior and biology.

TAKE-HOME POINTS

My goal is to have slightly altered your schemata. I hope that by this point you agree that we have busted, seriously challenged, or at least made much more complicated, the assumptions about human nature in relation to race, aggression, and sex.

Here are eight take-home points:

*Point 1: Humans are simultaneously biological and cultural;
we are naturenurtural creatures with a fascinating evolutionary
past and present*

We have complex biology (our bodies) and culture (our schemata) that create and shape, and are shaped by, our perceptions and philosophies. We have not evolved to have one specific way of being, and thus there are a number of potential outcomes to being human. However, our bodies, schemata, and cultural inheritance constrain the ways in which we can envision, construct, and experience those potential paths. We are born into a world of existing social and physical ecologies, patterns, and contexts that immediately become entangled with our biological structures and become a central part of our process of biocultural development. It is abundantly clear that simple explanations for who we are and why we do what we do are usually wrong.

Point 2: Culture matters

Our culture is a large part of what makes us unique. We can think of culture as a dynamic web of significance at the core of our becoming and being human. Culture is both a product of human action and something that influences that action; it is the context that helps give meaning to our experiences of the world. This means that cultural constructs are real for those that share them. Some constructs are more pervasive than others, and thus more important to understand because they affect how we live and act and treat others.

Point 3: Evolution matters

Evolution is largely misunderstood and misused when thinking about human behavior. Generally, evolution is change over time. Specifically, it is change in genotype and phenotype across generations due to a variety of processes. These processes create, move, and shape biological variation in multiple ways. It is not all about fighting and survival, it is about interactions between organisms and environments and organisms and each other over time. Niche construction theory shows us that humans and their environments are mutually interactive in evolutionary processes, and helps us realize that social and ecological inheritance is very important. Multiple inheritance theory illustrates that evolutionarily relevant inheritance can be genetic, epigenetic, behavioral, and symbolic. Evolution is not goal oriented nor does it produce endpoints: evolution is ongoing.

Point 4: Genes do not equal human nature

Our DNA alone does not determine who we are and how we behave, but it is a primary component in the development and maintenance of our bodies and behaviors. Genes contain the basic instructions for the building blocks (proteins) of biological systems. Genes and our phenotype (our bodies and behavior) are connected, but not usually in a one-to-one relationship; however, the relationship they do have is shaped and influenced by evolutionary processes, ecological and social contexts, and histories.

Point 5: Race is not what we think it is

The construct of human races is not a biological reality. There is substantial variation among individuals within populations and some biological variation is divided up between different populations and also among larger population groupings. Patterns of variation both within and between groups have been substantially shaped by culture, language, ecology, history, and geography. Race is not an accurate or productive way to describe modern human biological variation, but human variation research does have important social, biomedical, and forensic implications. Race is a cultural construct that can affect our social realities. Racial inequality (racism) is a social reality and can affect individuals' biology. Ethnicity is a valid way to ask questions about social histories and social and symbolic identification, but it is not biology and it is not race.

Point 6: Humans are not aggressive by nature

Humans have great potential for aggression and violence. There is variation in conflict styles and aggression across individuals, sexes, genders, societies, and time frames. Aggression itself is not a uniformly or consistently discrete trait, so aggression per se cannot be favored by evolutionary pressures. War is common in the human experience today, but it is not part of our evolutionary heritage. Human aggression, especially in males, is not an evolutionary adaptation: we are not aggressive, big-brained apes. Males and females differ in some facets of aggression, and many of those differences have to do with physical size and the social and experiential contexts in which the genders find themselves. Genes do not control or determine the normative expression of aggression, but abnormal biological function can influence aggressive behavior. Humans are hyper-cooperators and not "naturally" selfish. As a species we do not rely on aggression and violence more than cooperation and there is no pattern of evidence to support a notion that humanity is aggressive and selfish by nature. Aggression is an important part of being human, but it is not who we are at our core.

Point 7: Men and women are not as different as you may think

There are important biological differences between the sexes and there are also important similarities; however, there is a greater range of overlap in male and female bodies than most people realize. Behaviorally males and females also overlap extensively. Humans, regardless of sex, seek to form pair bonds of both social and sexual sorts, but pair bonds and marriage are not the same thing. Males and females, given the opportunity, will engage in sex across their lifespan in more or less the same rates and manners. These strong similarities in male and female bodies and behavior do not mean that gender differences are not real and important. Gender is a powerful cultural construct and the perception and expectation of gender differences impacts individuals and society. Males tend to control economic and political resources and women are heavily involved with child rearing because they give birth and lactate, but males and females have the same behavioral ability to care for offspring. There is no biological or evolutionary mandate that only females care for young and only males care for economics and politics. These patterns of gender difference and the strength of the cultural assumptions about sex fool us into thinking that men and women are different by nature.

Point 8: Busting myths about human nature requires critical thinking and a lot of work

Myths matter in our daily lives and make sense to our shared schemata, which is why they are tough to challenge. The information to bust most myths about humanity is largely available, but it exists across a range of academic disciplines, books, journals, media sources, and people. To tackle any of the myths about human nature one must compile a variety of information from different sources. Any single approach is not going to get you a sufficient set of information to achieve quality answers nor will it enable you to integrate the kinds of datasets needed to truly bust powerful stories about why we are the way that we are. Myth busting can alter the way we think about ourselves and the society around us; changing our minds is always a difficult and sometimes scary thing to do, but it is important.

WHY WERE WE LIED TO?

A lie is generally defined as a false statement made with deliberate intent to deceive; an intentional untruth; a falsehood, something intended or serving to convey a false impression; imposture; or an inaccurate or false statement.[1] Much in the common myths about human nature is a lie. The popular media, advertisements, Web sites, even academic publications can contain large amounts of misinformation and falsehoods associated with the basic ideas about race, aggression, and sex. I like to think that in most cases these lies are not on purpose. That is, the individuals arguing that men are from Mars and women are from Venus, that we are savage beasts by nature, and that race is underpinned by biology, actually believe what they are saying. They may believe it because it is a core part of their schemata or they might be doing research that provides some support for one part of the myth or they might simply be reading a series of books, articles, and Web sites that all push the same concept and thus are convinced. Schemata, inertia, historical precedence, and ignorance are very powerful: it is often difficult to challenge the status quo.

While not being active lying, there is a pattern in the scientific literature that acts to maintain some myths in roughly the same manner. It turns out that results that are exciting or show dramatic differences (like between males and females) tend to get more attention than do those studies that show small changes, that do not support hypotheses, or that show differences that are smaller than expected. This means that an idea that is new and/or supports a popular perspective can be widely repeated and incorporated as a baseline in other studies and projects, possibly without sufficiently rigorous analyses. In a 2002 study comparing major trends in the publication of research findings in evolution and ecology over time the biologists Michael D. Jennions and Anders P. Møller found that major conclusions of many studies were less and less supported as time passed from their first publication. That is, a big find is published and then a flurry of activity based on those findings ensues, but over time, with repeated testing, the research "fact" turns out to be less dramatic or significant than originally thought. However, this pattern of refutation takes a long time to gain widespread recognition, and might generally be ignored by much of the research establishment. This can be attributable to what is called the bandwagon effect or to the fact that there is a "publication bias against non-significant or weaker findings . . . non-significant results may take longer to publish."[2] This suggests that even scientists are influenced by their expectations and excitement over results, even when expanded testing demonstrates that the results are not as powerful as initially thought. This is extremely common in studies of issues related to humans (especially with sex and aggression). For example, the concept of symmetry-asymmetry being a major part of the biology of sexual attraction was extremely popular in the 1990s, but over time multiple studies demonstrated that its effect was much smaller or less clear than was initially proposed. However, this has not stopped many researchers, and the popular press, from assuming that it is still a major factor in the ways mates assess one another.[3]

However, there is a more nefarious reality in some cases of adamant supporters of these myths. Think about the arguments against racial or gender equality over the past few centuries that have been pushed by scientists, politicians, and even religious leaders with a specific intent: to support particular myths of difference by showing so-called scientific, and/or theological proof of superiority/inferiority. This involves manipulating actual information and data, even to the point of outright lying. People with particular biases and agendas can alter or selectively manipulate information and beliefs to make sure they support the appropriate outcomes.[4] This can then cascade into legal and social issues. Think about a lawyer using the warrior gene defense or the notion of the evolved nature of human male aggression to defend a client who committed violent rape, or certain individuals in the armed forces arguing that female reproductive frailty makes women unsuited for combat roles. Folks with this perspective and goal will select only the few scientists or experts who support that view and ignore the alternative possibilities or other voices.

This same pattern can be seen, to an extent, in our broader popular culture. Because the basic concepts in these myths are acquired and shared by most members of our society, they act to reproduce themselves at a subconscious level. They offer a common set of knowledge that can be exploited by advertisers, politicians, and other groups to reach a maximal audience. Reinforcing these myths acts as a self-fulfilling feedback loop, making them more robust. Examine the next ten television commercials you watch and see how many of them have a specific reference to, or a reliance on, some aspect of the myths of sex/gender and aggression; I bet it will be a majority of them. Interestingly, today (circa 2012) you will not see as many overt references to the myth of race in advertisements or in general popular media as this is a myth that is undergoing change and it is one that the public has lower receptivity for. There are also many legal structures in place to restrict overt public racism or racial discrimination. However, in society at large this myth still has a prominent impact.

Another important reason why misinformation, and even overt lies, remain prominent in the public mind is simplicity versus complexity in explanations. To counter what we think of, and experience, as everyday reality requires deconstruction of the assumptions, assessment of the assumptions, and then refutation of the assumptions—whereas supporting popular myths simply relies on reinforcing what you already "know," thanks to your schemata. Each of the myth-busting chapters in this book is around 20,000 words long and yet barely skims the surface of all the data and information available on the subject! Because of their powerful place in our culture these myths take much fewer words and ideas to be supported. Supporters of these myths can simply refer to common knowledge, to the world you "see" every day, and to many of the social experiences you yourself have had. They can ignore contradictory data, disregard variation, and focus on averages and cultural constructs, simply by ignoring the actual details of the issues involved. This works because of what we know about human biocultural development and the power of social context. These myths of human nature are promulgated so thoroughly, and successfully, in our society in large part because it is easier to accept the world as we perceive it is than to question and investigate whether the world might be a bit different than we think.

Finally, a core reason that these myths are so resilient is because few of us have the opportunity, the time, or the training to assemble the diverse array of information needed to effectively bust them. Remember the flat earth example from chapter 4: I am sure that the majority of you reading this book believe that the earth is round (more or less), and you are right. However, none of you have actually ever seen, for yourself, the curvature of the planet.[5] Five centuries ago most people believed the earth was flat; it was a core part of their schemata and a basic fact of the world. But since then we have been exposed to enormous amounts of evidence from images, to scientific data and publications, to popular stories and movies, and most importantly, you were taught in school that the earth is round. The flat earth myth was busted by thousands of people putting information into the public sphere and others collecting and synthesizing it until eventually the myth was exposed and the new fact of the shape of the world came into being. This is what needs to happen with the myths of race, aggression, and sex, but we are a long way from accomplishing it. This does not mean everyone should not have access to the basic tools of lie detection and myth busting to start the process themselves.

WHY SHOULD WE CARE?

The journalist Elton James White recently stated that "America talks about race like scared parents talk with their kids about sex. We're vague, sometimes terribly misleading and on occasion leave out huge aspects of the situation that would allow kids to make better decisions about how they conduct themselves. If we continue with our horrendously skewed and willfully ignorant interpretations of history, we will find ourselves with a generation that's woefully misinformed and it will be completely our fault."[6] We as a society do a poor job in really thinking about these issues, but we encounter them every single day. It matters how we think, and talk, about race, aggression, and sex.

Avoiding a head-on discussion or a serious challenge of our assumptions about race, aggression, and sex can impact our health, education, politics, economics, and gender relations.[7] You should care because you are part of a society and you are human. Your experiences and daily life are intertwined with those of everyone you interact with (as well as many you do not).

Realizing that white, black, and Asian are not real biological, evolutionary, or natural categories and that they do not reflect divisions in human nature allows us to make better sense of behavior and biology and move beyond inherent racism as an explanation for the world. However, understanding that race as a social construct affects biology, experience, and social context challenges many assumptions about inequalities in social status, educational attainment, and health which can be better addressed and ameliorated knowing that an underlying nature is not the explanation.

Knowing that there is variation in aggression across individuals, sexes, genders, societies, and time frames and that aggression itself is not a single unitary thing tells us that simple explanations about the innate nature of aggression are wrong. It helps us ask better questions about why and how violence occurs, which might give us better options for addressing it. Discarding the notion of an inevitability of male aggression and the concept that humans are innately aggressive and selfish forces us to see aggression and violence as a outcome of multiple contexts but not inevitable or a biological fallback in times of chaos or reduced social control. This makes it more difficult to find good answers, but it also makes it more possible to find true answers and solutions to the crises brought about by aggression and violence.

Knowing that males and females are not biological opposites or socially distinct units of humankind and that the genders overlap as much, if not more, than they differ helps all of us be ourselves. Challenging common assumptions about the battle between the sexes with this view enables society to accept the actual wide range of variation in gender, sexuality, and biology that characterizes humans. Knowing that we are not biologically preordained to be monogamously married and that the nuclear family is not necessarily the only proper biological goal for adults can also help us cope with the reality of variation in human behavior. At the same time, realizing the power of cultural expectations of gender, marriage, and sex and their interactions with our biology also forces us to be aware of the complexities of being human. It offers a wider range of insights when dealing with psychological and sexual problems than simply falling back on the perspective that men and women are from different planets.

Earlier in the book I stated that busting myths of human nature is not like busting a myth about the quality of air on airplane or the effectiveness of vitamin C at preventing a cold. I am sure that you agree. We cannot just do a quick test and show that such a myth, in its entirety, is flat out wrong. I warned you that there are very few easy tests and no easy explanations. As I said in chapter 1 busting myths about human nature means breaking the stranglehold of simplicity in our view of nature and forces us to realize that being human is very complicated. It means challenging common assumptions and a reliance on averages and popular perception and instead actually delving into the gritty details of what we know about what humans are made of and what they actually do. Now that you've read this book, I am hoping that you will agree that the basic myths about race, aggression, and sex are neither correct nor a core part of human nature. Being human is much more complex and much more interesting.

Getting the Information Yourself

So, you want to bust some popular myths? To effectively bust myths you need curiosity, perseverance, and the ability to collect and assess copious amounts of information. You have to be ready to read a lot, to think critically about what you read, and to be vigilant about popular assertions, whether you agree with them or not. Today it is also critical to have a computer and access to the Internet. The following are a few tips to think about as you consider trying to figure out what the world is actually like.

CHALLENGE COMMON SENSE!

Do not passively accept what you are told. If something strikes you as incorrect, or even if you just want to know if it is right or wrong, you need to do a little legwork to figure out how to assess the assertion. The easiest way to try this is to ask yourself the following questions about something you believe. Why do you believe it? Experience? School? Read it somewhere? Heard it from parents, a friend, a co-worker? First figure out where information comes from, then ask yourself why it feels right or wrong.

Sometimes your gut feeling reflects your schemata, telling you that all is as it should be. But sometimes you might, even subconsciously, have picked up on some of the variation, patterns, or inconsistencies that make you feel differently about this particular topic (whatever it is). This feeling might make you question some assertion of reality that seems clear and correct to others. How do you figure out how to effectively question the assertion? Clarifying what is actually being said and who is saying it is a good place to start. The assertion that boys play rougher than girls it is probably going to sound right, but how does anyone actually know this? If the person making the assertion does not study this then they must have heard it or read it somewhere and are simply repeating it. Or maybe the person making the assertion is going solely on their own life experience. Whatever the case, you need to figure out where the original information comes from (or at least where that person heard it) and start there.

For example, it makes common sense that men and women's brains are different. I challenged this in chapter 6. If you do not agree with my presentation of the data and conclusions in chapter 6 about male and female brains being so similar, then go and check it out for yourself. Check the notes to see the data and/or publications I am citing to support my points. Read them and see if you agree. Do a little research in libraries or on the Web, ask people, see where this path of inquiry takes you. In the end you might reaffirm your beliefs or maybe they might be challenged.

In this effort it is absolutely necessary that you are not afraid to admit you do not understand. We humans have very large and complex brains and one of their major functions is to help us identify when things just do not make sense to us. Being confused or not following a line of argument is a normal part of brain function, especially when it involves thinking about something that contradicts our normative expectations. Our society makes it difficult to admit we do not understand, but it is a critical first step in learning and in busting myths.

LOTS OF DATA

To integrate information across different areas of knowledge requires two steps. First, how do you get the information, and second, once you have it how can you tell what is most worthwhile and relevant? This can be broken down into three parts: Where do we find the information? How do we select the right information from the vast amount out there? How do we understand details of information in fields/areas that we know little or nothing about?

Where do we find the information? Most people know the answer to this question. Libraries, books, journals, print and visual media, the Internet, real discussions with real people, and professional scientific organizations are a good start. This involves what many folks in the field of education like to call life-long learning, which is the idea that the basic skills of searching for and acquiring information that you learn in school can be applied throughout life to keep enhancing and expanding your knowledge set.

As a baseline I encourage purchasing books (or e-books). The ability to own many, many books and therefore have continuous access to the knowledge they contain is a wonderful luxury of the modern age. The importance and value of reading for training the mind cannot be underestimated. However, when trying to do research in regard to myths of human nature, very few people are going to be able to spend the money to purchase hundreds of books. So I'll focus on the variety of sources that provide low-cost, or free, access to information.

If one has access to a university library this is often the best place to get a wide range of information. Today's university libraries have large holdings of books sorted by subjects, topics, and authors, but more importantly they also now have enormous digital holdings. Whether it is the tens of thousands of electronic PhD dissertations, the Web of science, Promed and other professional search databases, huge digital holdings of academic journals (some dating from the late 1800s), or vast online repositories of datasets, electronic access has become amazingly deep.[1] University libraries are treasure chests of access to information, but not everyone can use them. Public libraries vary dramatically. Some have excellent facilities and others do not. Some have great book hold-ings and electronic resources and others do not. It is important to see what your local public libraries have and to use them when possible, but they are much more variable in holdings and infrastructure support than are major university libraries.

There are a number of Internet sites that allow you to see a wide array of books for free. Google Books allows you to read sections of thousands of books for free. The online books page links you to a million free books via the Web.[2] Other sites, such as Darwin Online, allow you to read nearly everything Charles Darwin ever published (and much he did not publish).[3] The directory of open access journals (DOAJ) connects you with nearly 6,000 academic journals and almost 500,000 articles that are available to anyone for free.[4] Most academic journals today have Web sites where you can purchase access to individual articles as well. Finally, news Web sites and online newspapers can be good jumping-off points for ideas but not as sources of original data and analyses. You have to spend some time familiarizing yourself with all of the options, but there is a great deal of information at your fingertips in libraries, news media, and on the Web.

Given that there is so much out there, how do we select the right informa-tion from the vast amount available? Obviously, not everything (or even most things) on the Web are accurate. A Google search for "human nature" gives over 44 million results, most of which are not well researched or documented discussions on the topic. There are some general guidelines you can use to assess the validity of information on the Web.

First, for reliable data and assessments of those data you want to use sites that present information that is controlled via peer review or some other sort of rigorous oversight. University and research institute sites, governmental sites (like the Centers for Disease Control), and professional organizations often have high-quality suggested readings and links to basic information about their fields on their home pages.[5] Private and public foundations can have good sites, but be sure to carefully assess their missions and financial supporters to note what types of biases might be present. Wikipedia can be a decent tool for locating initial information but it is not yet reliable as a primary source of data and analyses. Google Scholar is becoming a relatively good clearinghouse for finding academic articles as well. Private Web sites set up by individuals might be reliable, but there is no quality control, assessment, or oversight, so always treat these sources with a certain amount of skepticism. The best types of private sites are usually good blogs that document where their information comes from and then connect the readers to original research that inspired the blogs and stories.[6]

Generally, science writers published in major newspapers and magazines are good sources for brief summaries of current issues related to aspects of human nature. However, these journalists have varying degrees of expertise in the areas they cover and often do not have a degree in the sciences. This means that they rely heavily on specific sources for their information and tend to present limited views of complex academic debates or research programs. They, like Wikipedia, are good jumping-off points. One way to tell the quality of science writing in the popular press is to see if links or references are provided to the actual study or studies being used to buttress the review's points.

Once you are able to access information you need to establish if it is from an original publication such as a peer-reviewed journal or Web site, or if it is an academic or popular book, or if the information came from a similar valid source.[7] If not, then you need to be skeptical of that information unless there is a clear description of how the information and conclusions presented were reached (such as good footnotes or bibliography, or other clear method of providing citations for where the information comes from). You also probably want to view multiple sites and locations of information and see if patterns emerge. These patterns can be cues that many people are finding/thinking the same thing, which can be an important context for your search, helping you assess the validity of the information.

However, when you do find academic articles and books or online sources you may notice that they can be highly specialized in biology or psychology or some other field that you might have no training in at all. How do you attempt to understand details in fields that you know little or nothing about? If you have access to introductory classes in the topic (either at a college or online) that is a terrific way to keep learning and expanding your knowledge set. If this is not an option you can purchase an up-to-date introductory textbook on the topic and read it, or look in bookstores for concise introductions or overviews of the topic.[8] It takes time to brush up on a topic, but in most cases with a bit of serious time and attention our big brains can pick up enormous amounts of information. Finally, a wide range of public lectures at museums, libraries, and universities, documentaries, TED lectures, and other sources of publicly available information are often enjoyable experiences and worthwhile ways to expand your knowledge base.[9]

Go for it!

Notes

CHAPTER 1

1. See chapter 4 of C. Geertz (1983), *Local Knowledge: Further Essays in Interpretive Anthropology.*

2. Mary Midgley (2004), *The Myths We Live By.*

3. See D.L. Hart and R.W. Sussman (2008), *Man the Hunted: Primates, Predators, and Human Evolution,* for an in-depth analysis of the myth of human ancestors and hunting. See also J.M. Adovasio, S. Olga, and P. Jake (2007), *The Invisible Sex: Uncovering the True Roles of Women in Prehistory,* for an overview of the ways in which we reconstruct gender roles in human evolutionary studies.

4. See chapter 6 and also Janet Shibley Hyde (2005), The gender similarities hypothesis, *American Psychologist* 60(6): 581–592.

5. See K.B. Strier (1994), Myth of the typical primate, *Yearbook of Physical Anthropology* 37:233–71, for a great example of this from the world of primate studies.

6. See Michael Shermer (2007), Airborne baloney, *Scientific American* (January) 296(1): 32.

7. See Patrick Smith's "Ask the Pilot" column March 7, 2007 for an excellent overview and links regarding this information; http://www.salon.com/technology/ask_the_pilot/2007/03/09/askthepilot224.

8. Comment by Dutch philosopher Fons Edlers, opening the 1971 debate between Michel Foucault and Noam Chomsky on human nature; see N. Chomsky and M. Foucault (2006), *The Chomsky-Foucault Debate: On Human Nature.*

9. Thomas Hobbes, *Leviathan,* chap. 13.

10. Comment by Michel Foucault in the 1971 debate between Michel Foucault and Noam Chomsky on human nature; see N. Chomsky and M. Foucault (2006), *The Chomsky-Foucault Debate: On Human Nature*.

11. Alan Miller and Satoshi Kanazawa (2007), Ten politically incorrect truths about human nature, *Psychology Today* (July 1), 40(4): 88–89.

12. Paul Ehrlich, from an interview with Natalie Angier, *New York Times*, October 10, 2000.

13. See Wikipedia.org/wiki/Human.nature; accessed November 2, 2011.

14. The band Love and Rockets, lyrics from the song "No new tale to tell."

15. The singer Madonna, lyrics from the song "Express yourself."

16. In 2009 an Italian court reduced the sentence of a convicted murderer because he tested positive for one version of a particular gene (MAOA) that is associated with some types of aggressive and loss of control behaviors (see *Scientific American*, October 30, 2009). Some people with this version of the gene seem to be more aggressive and less able to respond well to stress. The MAOA gene complex is touted as an example of male violent nature (see chapter 5).

17. See for example Dawkin's books *The Selfish Gene* (1976) and *The Extended Phenotype* (1982).

18. Thomas Aquinas (1873), *Summa Theologiae*. See also Craig A. Boyd (2004), Was Thomas Aquinas a sociobiologist? Thomistic natural law, rational goods, and sociobiology, *Zygon* 39(3): 659–680.

19. Okay, this is my coined word, and the best way I've come up with so far to demonstrate that nature and nurture really are not two separate things.

20. Poem can be read in its entirety at the University of Oxford's Thomas Gray Archive; http://www.thomasgray.org/cgi-bin/display.cgi?text=odec.

21. Here by "we" I primarily mean the United States, but this can be extended to a wide swath of the developed world.

22. From *The Times* (London), October 17, 2007; http://www.timesonline.co.uk/tol/news/uk/article2677098.ece.

23. Jonathan Marks (2009), *Why I Am Not a Scientist*, pp. 2–3.

24. For information on HIV and AIDS, see the Center for Disease Control's HIV Web site; http://www.cdc.gov/hiv/topics/basic/index.htm; also see http://www.avert.org/origin-aids-hiv.htm for brief overviews of a few of the different ideas about the origin of the AIDS virus, such as the oral polio vaccine concept, the colonialism concept, the contaminated needle concept, and the US government conspiracy concept.

25. See O.B. Frothingham (1900), Real case of the remonstrants against woman suffrage, *Arena* 2 (June): 175–81, in Edith M. Phelps, ed., *Debaters Handbook Series: Selected Articles on Woman Suffrage* (Minneapolis: N.W. Wilson Co., 1910).

26. See for example Deborah L. Rotman (2009), *Gendered Lives: Historical Archaeology of Social Relations in Deerfield, Massachusetts, ca. 1750–ca. 1904.*

27. For overviews see George J. Armelagos (2005), The slavery hypothesis—natural selection and scientific investigation: A commentary, *Transforming Anthropology* 13(2): 119–24; Lorena Madrigal et al. (2008), An evaluation of a genetic deterministic explanation for hypertension prevalence rate

inequalities, in C. Panter-Brick and A. Fuentes, eds., *Health, Risk, and Adversity,* pp. 236–55; and Jay Kaufman (2006), The anatomy of a medical myth, http://raceandgenomics.ssrc.org/Kaufman.

CHAPTER 2

1. Phyllis Dolhinow, a pioneer in American primatology and one of the first women to conduct primatological fieldwork (even before Jane Goodall), is emeritus professor of anthropology at the University of California at Berkeley. The quote is sometimes attributed to the Canadian intellectual Marshall McLuhan and is also the title of a 1975 publication by G.G. Foster, J.E. Ysseldyke, and J.H. Reese in the psychological journal *Exceptional Children.*

2. In fact, Karen Strier has an extensive and excellent article on this pattern in primatology; see K.B. Strier (1994), Myth of the typical primate, *Yearbook of Physical Anthropology* 37: 233–71.

3. The term "cognitive" as used here is an excellent example of what I mean by naturenurtural. Cognition reflects a true integration of our physiologies (brain, nervous system, and all their parts) and our life experiences (social events, cultural context, etc.) as the two shape one another throughout our lifetimes.

4. The term "biocultural" is gaining use among anthropologists who study human behavior, health, and evolution from an integrated perspective. It generally reflects attempts to examine the synthesis of biological and social factors without necessarily treating them as distinct things. Speaking from the biocultural perspective, Harold Odden states that "human processes commonly involve the active intersection and interaction of physiological, psychological, and sociocultural processes as well as larger macrolevel forces, such as poverty and inequality"; Harold L. Odden (2010), Interactions of temperament and culture: The organization of diversity in Samoan infancy, *Ethos* 37(2): 161–180. See also Alan H. Goodman and Thomas L. Leatherman, eds. (1998), *Building A New Biocultural Synthesis;* Catherine Panter-Brick and Agustín Fuentes, eds. (2008), *Health, Risk and Adversity;* Darna L. Dufour (2006), The 23rd annual Raymond Pearl Memorial Lecture. Biocultural approaches in human biology, *American Journal of Human Biology* 18(1): 1–9; and D.J. Hruschka, D. Lende, and C.M. Worthman (2005), Biocultural dialogues: Biology and culture in psychological anthropology, *Ethos* 33(1): 1–19.

5. From Tim Ingold (2000), *The Perception of the Environment: Essays on Livelihood, Dwelling and Skill,* p. 25.

6. Definitions for "schemata" include "diagrammatic presentation," or more broadly, "a structured framework or plan" or "outline"; and more specifically to the term's usage in anthropology, "a mental codification of experience that includes a particular organized way of perceiving cognitively and responding to a complex situation or set of stimuli."

7. Tim Ingold (2000), *The Perception of the Environment: Essays on Livelihood, Dwelling and Skill,* p. 25; see also Tim Ingold (1996), Situating action V: The history and evolution of bodily skills, *Ecological Psychology* 8 (2): 171–82. Ingold is not the first to present these ideas, but his synthesis of others' works

and his integration of these ideas into a broader anthropological context is masterful. Major components of these ideas come from such renowned theorists as Marcel Mauss, Claude Levi-Strauss, Gregory Bateson, and Pierre Bourdieu; also see the recent work by the anthropologist Greg Downey.

8. For further elaboration on this topic see Sarah Hrdy (2009), *Mother and Others: The Evolutionary Origins of Mutual Understanding.*

9. Of course relationship to heat is also affected by your biological baseline (whether you come from a long lineage of people from hot areas) and your social context (whether you are wealthy enough to have air conditioning on all the time).

10. See for example G. Rizzolatti and C. Sinigaglia (2008), *Mirrors in the Brain. How We Share Our Actions and Emotions;* Philip L. Jackson, Andrew N. Meltzoff, and Jean Decety (2006), Neural circuits involved in imitation and perspective-taking, *NeuroImage* 31(1): 429–39; and Thalia Wheatley, Shawn C. Milleville, and Alex Martin (2007), Understanding animate agents: Distinct roles for the social network and mirror system, *Psychological Science* 18: 469.

11. Or at least the ways in which a society's history is represented. History, rather than being a static entity, is a dynamic field, with the way we see our histories changes depending on who writes, speaks about, or teaches about them.

12. Actually, originally in the United States these were just for citizens who were of certain types of European descent and male, but currently we extend them to all humans (at least as an ideology). In fact, this set of ideals has been popular in Europe since the eighteenth century and has become fairly widespread since the mid-twentieth century with the establishment of the Universal Declaration of Human rights, which begins "Whereas recognition of the inherent dignity and of the equal and inalienable rights of all members of the human family is the foundation of freedom, justice and peace in the world"; http://www.un.org/en/documents/udhr/index.shtml.

13. Thomas Jefferson is, after all, the person often credited with the phrase "separation of church and state." See the March 12, 2010, *New York Times* article for a summary of the decisions; http://www.nytimes.com/2010/03/13/education/13texas.html.

14. Interestingly, this mimics our genetic system as well. As discussed later in this chapter and in chapter 3, humans all share the same genes, but vary in the forms of those genes. We all have the same general DNA, but are each genetically unique.

15. E.B. Tylor (1871), *Primitive Culture,* p. 1.

16. Franz Boas (1930), Anthropology, in E.A. Seligman and A. Johnson, eds., *Encyclopedia of the Social Sciences,* vol. 2, p. 79.

17. Alfred Kroeber and Clyde Kluckhohn (1952), Culture: A critical review of concepts and definitions. *Papers of the Peabody Museum, Harvard University,* 47, p. 357. Along with Franz Boas, Alfred Kroeber is widely credited with being one of the most important founding individuals in American anthropology.

18. Among anthropologists and other social scientists there is a vigorous debate as to what culture is and how we should use it. There are numerous books and articles covering this debate, but we will bypass it here for the sake of brevity.

19. John Hartigan, Jr. (2010), *Race in the 21st Century: Ethnographic Approaches,* p. 10.

20. Clifford Geertz (1973), *The Interpretation of Cultures,* p. 5.

21. Richard A. Shweder (2001), Rethinking the object of anthropology and ending up where Kroeber and Kluckhohn began, *American Anthropologist* 103(2): 437–40.

22. Although arguments for the nuclear family are frequently made by invoking some specific connection between humans, primates, and monogamy and a nuclear family structure, this is not borne out by actual evolutionary or social data. See chapter 6.

23. In the United States people often confuse race and ethnicity. The two are not the same: ethnicity generally refers to a suite of shared historical, cultural, geographic, and linguistic features that are used to group peoples into clusters whereas race involves these characteristics but also has specific biological factors. The two terms are often used interchangeably, but technically this is an incorrect usage. More on this in chapter 4.

24. See these overview texts for fuller discussion and specific examples: John Hartigan, Jr. (2010), *Race in the 21st Century. Ethnographic Approaches;* J.L. Carroll (2010), *Sexuality Now: Embracing Diversity;* C.P. Kottack and K.A. Kozaitis (2003), *On Being Different: Diversity and Multiculturalism in the North American Mainstream;* and V.N. Parillo (2009), *Strangers to These Shores.*

CHAPTER 3

1. Nicolas Wade, Adventures in very recent evolution, *New York Times,* July 19, 2010.

2. It is interesting that Darwin did not use this term in the first edition of *On the Origin of Species,* the book where he lays out his proposal for natural selection. He inserted it into later editions at the urging of a few colleagues, but apparently was never fully comfortable with the phrase. His initial way of describing the processes of natural selection was "descent with modification" and he put a less emphatic focus on direct competition between organisms. See R. Richards (1987), *Darwin and the Emergence of Evolutionary Theories of Mind and Behavior.*

3. For a particularly robust and engaging overview of just how erroneous this view of competition as a driving force in evolution is see K.M. Weiss and A.V. Buchanan (2009), *The Mermaid's Tail: Four Million Years of Cooperation in the Making of Living Things.*

4. See for example R. Richards (1987), *Darwin and the Emergence of Evolutionary Theories of Mind and Behavior;* P. Ehrlich (1999), *Human Natures: Genes, Cultures and the Human Prospect;* K. Malik (2002), *Man, Beast, and Zombie: What Science Can and Cannot Tell Us About Human Nature;* and Stephen Jay Gould (2002), *The Structure of Evolutionary Theory.*

5. See two books by the journalist Jon Entine for these positions: *Taboo: Why Black Athletes Dominate Sports and Why We're Afraid to Talk About It* (2001) and *Abraham's Children: Race, Identity and the DNA of the Chosen*

People (2008) and also seek out the wide array of reviews of these books, such as those by Jon Marks (2000), *Human Biology* 72(6):1074–1078; Paul Achter and Celeste Condit (2000), *American Scientist* May-June (http://www .americanscientist.org/bookshelf/pub/not-so-black-and-white); and Harry Ostrer (2008), *Nature Genetics* 40(2): 127.

6. Visit the National Center for Science Education (http://ncse.com) for an excellent overview of the evidence and for great teaching tools, links to other sites, and essays about the reality of evolution.

7. It turns out that a few types of life, like some viruses, use a molecule very similar to DNA called RNA in place of DNA. All organisms that have DNA also have RNA, but some have only RNA. Also, a few things that act as if they are life forms, but we are not sure if they are, like prions that cause mad cow disease, do not have any DNA or RNA. However, these are the extreme cases and very rare. As a general rule the vast majority of life forms share DNA as a basic component of their structure and function.

8. James Watson and Francis Crick successfully described the structure of the DNA molecule in 1953: the DNA molecule is a double helix (usually described as a twisted ladder). Chemically, DNA is made up of three parts: nucleotide bases, sugars, and phosphates. A collection of sugars and phosphates are the backbone and two nucleotide bases make up each rung. Although we talk about DNA as ladder-like, it almost never actually appears in this form. Most of the time, DNA is tightly bound up and condensed into chromosomes, which are super-coiled masses of DNA in the nucleus of your cells. Human DNA is grouped onto 46 chromosomes, or 23 pairs of chromosomes. Each person has two copies of chromosomes 1 through 22, one from the mother and one from the father. Chromosome 23 is the pair of sex chromosomes and comes in two forms: an X and a Y. If you receive an X from your mother and an X from your father, your sex is female. If you have an X from your mother and a Y from your father, your sex is male (more on the sex chromosomes in chapter 6).

9. In most sexually reproducing organisms the female (mother) also provides a suite of important non-DNA elements in the egg. So you inherit DNA from both parents, but you inherit extra structural material from your mother.

10. In this usage a "population" is the collection of people (or organisms) that reside in more or less the same place, or in different places but are constantly connected, and mate more with one another than with members of other such populations.

11. In reality, genetics is much more complex. The definition of a gene and the enormous complexity of the genomic system (what people who actually work with genes now call genetics) are too much to delve into here. I've laid out the basics as you might find them in an introductory textbook, sufficient to use as a critical tool but nowhere near enough detail to dive in to real genetic analyses. For more background on genes see E. Fox-Keller (2000), *The Century of the Gene,* and N.C. Comfort (2001), Are genes real? *Natural History* (June), pp. 28–38. For a good overview of genetics, seek out an introductory genetics textbook or have a look at Tara R. Robinson (2010), *Genetics for Dummies.*

12. Agustín Fuentes (2011), *Biological Anthropology: Concepts and Connections.*

13. See for example P. Ehrlich and M. Feldman (2002), Genes and culture: What creates our behavioral phenomena? *Current Anthropology* 44(1): 87–107.

14. Major genetic mutations such as changes in the chromosomes or losses of large segments of DNA can alter the trajectory and patterns of growth of an embryo as can certain mutations in regulatory sequences. But these are not that common and usually result from dramatic exposure to external elements causing mutation such as radiation or toxins.

15. The phenotype is the observable and measurable traits of an organism, thus more or less everything above the level of the genes. An organism's phenotype is a collection of all of its traits.

16. Actually it was jointly proposed by Charles Darwin and Alfred Russel Wallace in 1858 in a paper to the Linnean Society of London (http://wallacefund.info/the-1858-darwin-wallace-paper), but Darwin's book *On the Origin of Species* was the full elaboration of the concept. Since that time countless researchers have expanded on, tested, and supported Darwin's and Wallace's basic premise. For an excellent overview see the UC Berkeley evolution site (http://evolution.berkeley.edu) or read the very good college textbook by Douglas Futuyma (2009), *Evolution,* 2nd ed.

17. Basically the most common current definition for evolution is changes in allele frequencies in populations across generations.

18. For an in-depth overview of niche construction see F. John Odling-Smee, Kevin N. Laland, and Marcus W. Feldman (2003), *Niche Construction: The Neglected Process in Evolution.* This concept draws heavily on earlier work by the biologists Richard Lewontin, Ernst Mayr, and Conrad Waddington as well. See Richard Lewontin (1983), Gene, organism and environment, in D.S. Bendall, ed., *Evolution from Molecules to Men;* Conrad H. Waddington (1959), Evolutionary adaptation, in S. Jax, ed., *Evolution after Darwin,* pp. 381–402, and E. Mayr (1963), *Animal Speciation and Evolution.*

19. Kevin Laland, Jeremy Kendall, and Gillian Brown (2007), The niche construction perspective: Implications for evolution and human behavior, *Evolutionary Psychology* (5): 51–66.

20. See Peter J. Richerson and Robert Boyd (2005), *Not by Genes Alone: How Culture Transformed Human Evolution,* for an excellent overview of the gene-culture coevolution ideas, and Kevin Laland, Jeremy Kendall, and Gillian Brown (2007), The niche construction perspective: Implications for evolution and human behavior, *Evolutionary Psychology* (5): 51–66 for a concise integration of niche construction ideas into this paradigm.

21. See Eva Jablonka and Marion Lamb (2005), *Evolution in Four Dimensions: Genetic, Epigenetic, Behavioral, and Symbolic Variation in the History of Life,* for an excellent overview of this perspective.

22. Epigenetic refers to physiological/biological processes in the organism at a level of organization above the DNA but still within the body proper. These can be cells, tissues, organs, muscles systems, neuronal systems, and so on.

PRELUDE

1. Theodosius Dobzhansky (1972), On the evolutionary uniqueness of man, in T. Dobzhansky, M.K. Hecht, and W.C. Steere, eds., *Evolutionary Biology,* vol. 6, p. 425.

2. The quote is attributed to Einstein, but it is not clear that he actually said this.

3. "Schemata" is anthropology's technical term for a person's comprehensive, acquired, and complex worldview (discussed in chapter 2) and "naturenurtural" is my suggestion for the best way to envision the full fusion of biology and culture rather than treating them as two distinct things that we simply add together to understand humans.

CHAPTER 4

1. Ashley Montagu (1942), *Man's Most Dangerous Myth: The Fallacy of Race.*

2. Guy Harrison (2010), *Race and Reality: What Everyone Should Know About Our Biological Diversity,* p. 20.

3. Remember the saying in chapter 2, "I would not have seen it if I hadn't believed it."

4. http://physanth.org/association/position-statements/biological-aspects-of-race/?searchterm=race. This statement is the official stance of the association of scientists (physical or biological anthropologists) who have spent the last 150 years examining human biological variation.

5. This quote is from J. Phillipe Rushton (2000), *Race, Evolution, Behavior: A Life History Perspective,* 2nd ed., p. 17. Rushton is a psychologist focused on "proving" the biological basis for race categories. His work has been reviewed, refuted, and rejected by anthropology and biology journals due to its lack of scientific support and his selective use of fringe datasets. He remains active in self-publishing and also still publishes in a few psychology journals. However, his pronouncements about race differences are worth reading as they receive wide attention and are used by many in the lay public (and racist groups) to shore up assertions about a biological basis for race.

6. There have been numerous refutations of Rushton's work. The American Association of Physical Anthropologists refused his membership request on the grounds of his consistent manipulation of information and his continued pushing of racist ideology in spite of countless refutations of his published works. See the rest of this chapter and the Web sites http://www.understandin-grace.org/home.html and http://raceandgenomics.ssrc.org for a whole series of examples and articles that deal with the assertions put forward by Ruston and others supporting the reality of biological races in modern humans.

7. Of course, there are five race systems that add Native American and Hispanic, and other systems that add even more. We address this later in the chapter. For this basic introduction we focus on the standard "big three" division.

8. Taxonomy is the science of naming or classifying organisms based on their phenotypes. Linnaeus developed the system of binomial nomenclature (two names) that we use today. His basic system lumps organisms based on

similarity to one another. See http://www.ucmp.berkeley.edu/history/linnaeus. html for a brief overview.

9. A subspecies, also referred to as a biological race, is a unit within a species that is taking an evolutionary path different from the overall trajectory of other populations within the species and thus is becoming increasingly differentiated from them at the genetic level. See A. Templeton (1999), Human races: A genetic and evolutionary perspective, *American Anthropologist* 100: 632–50 and R.A. Kittles and K.M. Weiss (2003), Race, genes and ancestry: Implications for defining disease risk, *Annual Reviews in Human Genetics* 4: 33–67 for overviews.

10. This selection (and translation) is from Jon Marks's 1995 excellent overview of Linnaeus and racial taxonomies (p. 50). I leave out *H. s. monstrosous* as this one was not based any actual peoples at all. See J. Marks (1995), *Human Biodiversity: Genes, Race, and History* for a fuller discussion of this topic.

11. Again see J. Marks (1995), *Human Biodiversity: Genes, Race, and History* for a fuller, and extremely engaging, discussion of this.

12. See for example K.M. Weiss (1998), Coming to terms with human variation, *Annual Reviews in Anthropology*, 27: 273–300; S. Molnar (2002), *Human Variation: Races, Types and Ethnic Groups*; J.H. Relethford (2002), Apportionment of global human genetic diversity based on craniometrics and skin color, *American Journal of Physical Anthropology*, 118: 393–98; N.G. Jablonski (2004), The evolution of human skin and skin color, *Annual Reviews in Anthropology*, 33: 585–623; C. Ruff (2002), Variation in human body size and shape, *Annual Reviews in Anthropology*, 31: 211–32; and A. Smedley and B. Smedley (2005), Race as biology is fiction, racism as a social problem is real: Anthropological and historical perspectives on the social construction of race, *American Psychologist* 60(1): 16–26. This is just a small sample of the published, peer-reviewed research and reviews that debunk the race classifications.

13. The circulatory system includes the blood, the blood vessels (arteries, capillaries, and veins), and the heart. This system is the core means of transport for the elements required by the body's tissues for survival.

14. These proteins and their relative compatibility are what set up the problems trying to transfer blood between individuals. In order to be able to transfer blood between humans their blood systems must be very compatible (at least the major ones—ABO and Rh).

15. When talking about alleles, when we say "dominant" we mean that when a dominant allele is paired with a recessive allele in the body only the protein product of the dominant allele is expressed. The recessive allele's protein is not produced, or is produced at very low levels so that the phenotype expressed is that influenced by the dominant alleles. Remember from the last chapter that these alleles are for the same gene and that gene comes in two copies per person.

16. When looking at blood group allele frequencies, it is good to look at indigenous populations or populations that have been in the same location for a long time to see the patterns of allele distribution; if we were to take a random sample from a place like New York City, it would reflect the species-wide averages because of all the gene flow.

17. See S. Molnar (2002), *Human Variation: Races, Types and Ethnic Groups* for a good overview of blood group variation in humans.

18. Interestingly, today this variation creates a problem: it makes organ transplants very difficult as our immune systems are so variable. Trying to transfer tissue from one human to another, even from one family member to another, can result in rejection of the transplanted tissue by the host body due to its different HLA protein structure.

19. See S. Molnar (2002), *Human Variation: Races, Types and Ethnic Groups* for a good overview.

20. For more on sickle cell see http://www.understandingrace.org/humvar/ sickle_01.html and http://www.nhlbi.nih.gov/health/dci/Diseases/Sca/SCA_ WhatIs.html.

21. See for example R.A. Kittles and K.M. Weiss (2003), Race, genes and ancestry: Implications for defining disease risk, *Annual Reviews in Humans Genetics* 4: 33–67; R.C. Lewontin (1972), The apportionment of human diversity, *Evolutionary Biology* 6: 381–98; J.C. Long, J. Li, and M.E. Healy (2009), Human DNA sequences: More variation and less race, *American Journal of Physical Anthropology* 139(1): 23–34; A. Templeton (1999), Human races: A genetic and evolutionary perspective, *American Anthropologist* 100: 632–50, K.M. Weiss (1998), Coming to terms with human variation, *Annual Reviews in Anthropology* 27: 273–300. For a basic statement on ancestry testing and genetic variation see D.C. Royal, J. Novembre, S.M. Fullerton, D.B. Goldstein, J.C. Long, M.J. Barnshad and A.G. Clark (2010), Inferring genetic ancestry: Opportunities, challenges, and implications, *American Journal of Human Genetics* 86: 661–73.

22. While sounding shocking at first, this is not really that surprising if you step back and consider that in our DNA all humans and all chimpanzees are about 98 percent identical. For that matter we also share about 40 percent of our DNA with daffodils. See Jon Marks (2002), *What It Means to Be 98% Chimpanzee.*

23. This type of comparison oversimplifies the ways in which human genetics varies across the species, but even the more complex ways of looking at genetic variation still clearly demonstrate that the our DNA does not support the concept of multiple human races today. See J.C. Long, J. Li, and M.E. Healy (2009), Human DNA sequences: More variation and less race, *American Journal of Physical Anthropology* 139(1): 23–34; R.A. Kittles and K.M. Weiss (2003), Race, genes and ancestry: Implications for defining disease risk, *Annual Reviews in Humans Genetics* 4: 33–67; and A. Templeton (1999), Human races: A genetic and evolutionary perspective, *American Anthropologist* 100: 632–50.

24. R.A. Kittles and K.M. Weiss (2003), Race, genes and ancestry: Implications for defining disease risk, *Annual Reviews in Human Genetics* 4: 33–67; A. Templeton (1999), Human races: A genetic and evolutionary perspective, *American Anthropologist* 100: 632–50.

25. D.C. Royal, J. Novembre, S.M. Fullerton, D.B. Goldstein, J.C. Long, M.J. Barnshad, and A.G. Clark (2010), Inferring genetic ancestry: Opportunities, challenges, and implications, *American Journal of Human Genetics* 86: 661–73.

26. The reference samples for this research are most commonly drawn from fifty-one populations living in different parts of the world; the database is

called the Human Genetic Diversity Panel (HGDP). However, as noted by the American Society of Human Genetics, "Although the HGDP collection is a useful collection of widely distributed human populations, it is a convenient sample and does not sample densely within any one geographic region; hence, there are limitations to the accuracy of ancestry inference within and among regions." In other words, relying on this database helps us get some general ideas about ancestry, but it is a small and incomplete version of human population variation and does not reflect racial divisions. See D.C. Royal, J. Novembre, S.M. Fullerton, D.B. Goldstein, J.C. Long, M.J. Barnshad, and A.G. Clark (2010), Inferring genetic ancestry: Opportunities, challenges, and implications, *American Journal of Human Genetics* 86: 661–73.

27. Here functional means the direct product of evolution via natural selection. That is, the variant is there because it has served as an adaptation in an evolutionary sense. This is not the case for AIMs.

28. In fact the study by the American Society of Human Genetics states these limitations best: "To infer ancestry, researchers rely on comparing any individual's particular genetic profile to that of reference populations. Research geneticists benefit from various publicly available databases such as the HapMap, Human Genome Diversity Panel, Perlegen Human Genome Resources, POPRES project, and Seattle SNPs projects. However, even the databases that researchers consider the most applicable reflect a woefully incomplete sampling of human genetic diversity, and this has important consequences for the accuracy of ancestry inference. One problem is that the 'ancestral populations' assumed by some methods are not explicitly represented in databases—and indeed cannot be represented as such because we do not have the ability to sample ancestral populations. A second problem is that populations that are mixtures of the 'typical' reference populations (e.g., Africans, Asians, and Europeans) are substantially under-represented in these databases. Recent sampling efforts, such as HapMap Phase III samples, are helping to remedy this problem; however, continued attention to diverse sampling will be an important aspect of any subsequent surveys of human genetic variation." D.C. Royal, J. Novembre, S.M. Fullerton, D.B. Goldstein, J.C. Long, M.J. Barnshad, and A.G. Clark (2010), Inferring genetic ancestry: Opportunities, challenges, and implications, *American Journal of Human Genetics* 86: 661–73.

29. J.C. Long, J. Li, and M.E. Healy (2009), Human DNA sequences: More variation and less race, *American Journal of Physical Anthropology* 139(1): 23–34; R.A. Kittles and K.M. Weiss (2003), Race, genes and ancestry: Implications for defining disease risk, *Annual Reviews in Humans Genetics* 4: 33–67; and A. Templeton (1999), Human races: A genetic and evolutionary perspective, *American Anthropologist* 100: 632–50.

30. See Agustín Fuentes (2011), *Biological Anthropology: Concepts and Connections,* 2nd ed.; Michael Alan Park (2009), *Biological Anthropology,* 6th ed., or any introductory biological anthropology or human evolution textbook for a concise overview of human evolutionary history and the central role of the African continent in human evolution.

31. C. Ruff (2002), Variation in human body size and shape, *Annual Reviews in Anthropology* 31: 211–32.

32. Read Nina Jablonski's *Skin: A Natural History* (2006) for the best and most comprehensive overview of everything you even wanted to know about the wonderful coverings of our bodies. What follows here is a very brief summary of the in-depth analyses provided by Jablonski.

33. See Jablonski's *Skin,* but also see core research articles such as N. Jablonski and G. Chaplin (2000), The evolution of human skin color, *Journal of Human Evolution* 39: 57–106; N.G. Jablonski (2004), The evolution of human skin and skin color, *Annual Reviews in Anthropology* 33: 585–623; and J.H. Relethford (2002), Apportionment of global human genetic diversity based on craniometrics and skin color, *American Journal of Physical Anthropology* 118: 393–98.

34. J.H. Relethford (2002), Apportionment of global human genetic diversity based on craniometrics and skin color, *American Journal of Physical Anthropology* 118: 393–98.

35. See R.L. Jantz and L. Meadows Jantz (2000), Secular change in craniofacial morphology, *American Journal of Human Biology* 12: 327–38 and S. Ousley, R. Janytz, and D. Freid (2009), Understanding race and human variation: Why forensic anthropologists are good at identifying race, *American Journal of Physical Anthropology* 139(1): 68–76.

36. One of the very few patterns that does fix largely to a geographic area is the thick, very straight hair that is found in high frequencies in northeastern and eastern Asia. While the hair type is found in populations outside of East Asia its high frequency is probably attributable to widespread gene flow in East Asia by peoples originating in northeastern Asia. Frizzy hair, associated exclusively with being "black" in the United States, is found in populations in Africa, South Asia, Southeast Asia, and Melanesia, and is not characteristic of all populations in Africa.

37. Yolanda Moses (2004), The continuing power of the concept of "race," *Anthropology and Education Quarterly* 35(1): 146–48.

38. This ranges from commentators on major media outlets (for example see http://www.salon.com/mwt/feature/2010/08/13/dr_laura_the_n_word) to Web sites greeting the postracial society, but at the same time polling indicates that nearly 50 percent of US citizens see racism as a major problem. However, blacks see this as a much greater issue than whites. See the overview of polls related to race in 1999–2009 at http://www.pollingreport.com/race.htm.

39. http://www.aaanet.org/issues/policy-advocacy/AAA-Statement-on-Race .cfm.

40. ABC News/Washington Post poll, January 13–16, 2009.

41. These are directly drawn from the definitions and guidelines for the 2010 census (http://2010.census.gov/partners/pdf/langfiles/qrb_English.pdf). As explained in http://www.understandingrace.org/about/response.html, "the Statistical Policy Division, Office of Information and Regulatory Affairs, of the Office of Management and Budget (OMB) determines federal standards for the reporting of 'racial' and 'ethnic' statistics. In this capacity, OMB promulgated Directive 15: Race and Ethnic Standards for Federal Statistics and Administrative Reporting in May, 1977, to standardize the collection of racial and ethnic information among federal agencies and to include data on persons

of Hispanic origins, as required by Congress. Directive 15 is used in the collection of information on 'racial' and 'ethnic' populations not only by federal agencies, but also, to be consistent with national information, by researchers, business, and industry as well."

42. http://2010.census.gov/partners/pdf/langfiles/qrb_English.pdf.

43. See the following reports and Web sites for data about racial disparities: Labor Force Characteristics by Race and Ethnicity (2009), U.S. Department of Labor U.S. Bureau of Labor Statistics November 20010 Report 1020; http://www.bls.gov/cps/cpsrace2009.pdf; http://www.msnbc.msn.com/id/16831909; http://www.kaiseredu.org/topics_reflib.asp?id=329&rID=3&parentid=67; and Pew Research Center Report on Social and Demographic trends (2011), http://pewsocialtrends.org/2011/07/26/wealth-gaps-rise-to-record-highs-between-whites-blacks-hispanics.

44. Troy Duster (2005), Race and reification in science, *Science* 307: 1050–51.

45. Race is by no means the only contributor to patterns of inequality in the United States. Ethnicity (more categories than race), annual income, religion, language, region of the country one lives in, level of education, and many other factors also go into making up the various patterns of Inequality we see across our society.

46. The preface to the Department of Health and Human Services publication, *Health, United States, 2009* explains that this is "the 33rd report on the health status of the Nation and is submitted by the Secretary of the Department of Health and Human Services to the President and the Congress of the United States in compliance with Section 308 of the Public Health Service Act. This report was compiled by the Centers for Disease Control and Prevention's (CDC) National Center for Health Statistics (NCHS). The National Committee on Vital and Health Statistics served in a review capacity. The *Health, United States* series presents national trends in health statistics. Each report includes an executive summary, highlights, a chart book, trend tables, extensive appendixes, and an index." For access, see http://www.cdc.gov/nchs/default.htm.

47. Clarence Gravlee (2009), How race becomes biology: Embodiment of social inequality, *American Journal of Physical Anthropology* 139: 47–48.

48. Note how the company uses the terms "Caucasian" and "African American" as opposed to white and black, even though they clearly mean the US categories of "white" and "black." It is interesting to consider why these ads do not use "European American" instead of "Caucasian," in parallel with "African American."

49. Actually, these data are misleading because they inflate the actual differences between blacks and whites as only 6 percent of heart failure (risk for hypertension) occurs in that age group, and in the above-65 age group there is no difference between whites and blacks; Troy Duster (2005), Race and reification in science, *Science* 307: 1050–51.

50. Debate continues as to whether this drug actually does act more effectively in blacks as opposed to other groups and why that might be. The original sample size was only forty-nine and the differences in effectiveness were very

small. See Jonathan Kahn (2007), Race in a bottle, *Scientific American* 297(2): 40–45.

51. See Lorena Madrigal, M. Blell, E. Ruiz, and F. Otárola-Durán (2008), An evaluation of a genetic deterministic explanation for hypertension prevalence rate inequalities, in C. Panter-Brick and A. Fuentes, eds., *Health, Risk, and Adversity*, pp. 236–55, and Troy Duster (2005), Race and reification in science, *Science* 307: 1050–51.

52. See W.W. Dressler, K.S. Oths, and C.C. Gravlee (2005), Race and ethnicity in public health research: Models to explain health disparities, *Annual Review of Anthropology* 34: 231–52; N. Krieger and S. Sidney (1996), Racial discrimination and blood pressure: The CARDIA study of young black and white adults, *American Journal of Public Health* 86(10): 1370–78; and N. Krieger, K. Smith. D. Naishadham, C. Hartman, and E.M. Barbeau (2005), Experiences of discrimination: Validity and reliability of a self-report measure for population health research on racism and health, *Social Science and Medicine* 61: 1576–96.

53. C.C. Gravlee, A.L. Non, C.J. Mulligan (2009), Genetic ancestry, social classification, and racial inequalities in blood pressure in southeastern Puerto Rico, *PLoS ONE* 4(9): e6821.

54. Peter B. Bach, Hoangmai H. Pham, Deborah Schrag, Ramsey C. Tate, and J. Lee Hargraves (2004), Primary care physicians who treat blacks and whites, *New England Journal of Medicine* 351: 6.

55. Lawrence Summers' (former president of Harvard University) remarks at the National Bureau of Economic Research Conference on Diversifying the Science and Engineering Workforce, January 14, 2005.

56. For example see John Hartigan (2010), *Race in the 21st Century: Ethnographic Approaches* for excellent summaries of these factors in thinking about "white" and "black" racial experiences and histories in the United States.

57. The Institute for Diversity and Ethics in Sports (TIDES) at the University of Central Florida, "The 2009 Racial and Gender Report Card: National Football League" (http://www.tidesport.org/RGRC/2011/2011_NFL_RGRC.pdf).

58. The Institute for Diversity and Ethics in Sports (TIDES) at the University of Central Florida, "The 2010 Racial and Gender Report Card: National Basketball Association" (http://www.tidesport.org/RGRC/2011/2011_NBA_RGRC[1].pdf).

59. The Institute for Diversity and Ethics in Sports (TIDES) at the University of Central Florida, "The 2010 Racial and Gender Report Card: Major League Baseball" (http://www.tidesport.org/RGRC/2011/2011_MLB_RGRC_updated.pdf). See also Steve Wulf (1987), "Standing tall at short" *Sports Illustrated*, February 9.

60. In particular see J. Marks (1995), *Human Biodiversity: Genes, Race, and History* and Guy Harrison (2010), *Race and Reality: What Everyone Should Know about Our Biological Diversity*. Also, visit http://www.understandingrace.org/lived/sports/index.html.

61. During the first forty years of the twentieth century, many southern and eastern European peoples (especially recent immigrants) were not lumped with US citizens of western and northern European descendance as "white." The

ethnicities of "Italian," "Jewish," "Romanian," and so on were seen as distinct biological races, much as "black" and "white" are today.

62. See Rushton's work as well as R. Hernnstein and C. Murray (1994), *The Bell Curve: Intelligence and Class Structure in American Life* or even Jon Entine (2000), *Taboo: Why Black Athletes Dominate Sports and Why We're Afraid to Talk about It,* which Entine would say is not eugenics oriented or influenced, but many others would argue is.

63. For an excellent discussion of this see Kenneth M. Weiss (2008), The good, the bad, and the ugli: Can seemingly incompatible worldviews be hybridized? *Evolutionary Anthropology* 17: 129–34.

64. Heather J.H. Edgar and Keith L. Hunley (2009), Race reconciled? How biological anthropologists view human variation, *American Journal of Physical Anthropology* 139: 1–4.

65. They Might Be Giants (1990) from their song "Your Racist Friend" on the album *Flood*.

CHAPTER 5

1. Except maybe for the interesting case of some chimpanzee aggression discussed later in this chapter.

2. R. Wrangham and D. Peterson (1996), *Demonic Males: Apes and the Origin of Human Violence,* pp. 22 23.

3. S. Pinker (2002), *The Blank Slate: The Modern Denial of Human Nature,* p. 316.

4. Azar Gat (2000), The causes and origins of "primitive warfare": Reply to Ferguson, *Anthropological Quarterly* 73(3): 165–68.

5. R. Wrangham and D. Peterson (1996), *Demonic Males: Apes and the Origin of Human Violence,* pp. 108–109.

6. L. Lehmann and M.W. Feldman (2008) War and the evolution of belligerence and bravery, *Proceedings of the Royal Society B* 275: 2877.

7. See Douglas Fry (2007), *Beyond War: The Human Potential for Peace* for a series of very good cross-cultural examples of how people typically interact peacefully.

8. J. Archer (2009), The nature of human aggression, *International Journal of Law and Psychiatry* 32: 202–208.

9. It is good to keep in mind that in the context of normal aggression there are many positive social outcomes. For example, consider the mild aggression used by parents (in humans and in other primates) when socializing young or the aggression that individuals within a group use when breaking up fights or policing and resolving conflicts between group members. See I.S. Bernstein (2007), Social mechanisms in the control of primate aggression, in C.J. Campbell, A. Fuentes, K.C. MacKinnon, M. Panger, and S.K. Bearder, eds., *Primates in Perspective,* pp. 562–70 and J.C. Flack, M. Girvan, F.B.M. de Waal, and D.C. Krakauer (2006), Policing stabilizes construction of social niches in primates, *Nature* 439: 426–29.

10. See R.J. Nelson and B.C. Trainor (2007), Neural mechanisms of aggression, *Nature Reviews: Neuroscience* 8: 536–46; A. Siegel and J. Victoroff (2009),

Understanding human aggression: New insights from neuroscience, *International Journal of Law and Psychiatry* 32: 209–15; P.R. Shaver and M. Mikulincer (2011), Introduction, in P.R. Shaver and M. Mikulincer, eds., *Human Aggression and Violence: Causes, Manifestations, and Consequences*, pp. 3–11; C.N. DeWall and C.G. Anderson (2011), The general aggression model, in P.R. Shaver and M. Mikulincer, eds., *Human Aggression and Violence: Causes, Manifestations, and Consequences*, pp. 15–33.

11. A. Siegel and J. Victoroff (2009), Understanding human aggression: New insights from neuroscience, *International Journal of Law and Psychiatry* 32: 209–215.

12. See D.L. Hart and R.W. Sussman (2008), *Man the Hunted: Primates, Predators, and Human Evolution*.

13. See A. Siegel and J. Victoroff (2009), Understanding human aggression: New insights from neuroscience, *International Journal of Law and Psychiatry* 32: 209–215 and J. Archer (2009b), The nature of human aggression, *International Journal of Law and Psychiatry* 32: 202–208.

14. It is worth noting here that the term "predatory" is drawn from the predator-prey relationship, largely because of its proactive nature as opposed to reactive, defensive aggression. In this chapter we are focusing on aggression within species, not the behaviors used in hunting and capturing prey items. It should be obvious that while the behaviors used in prey capture and in proactive aggression might be similar, the stimulus and reasons for intraspecific aggression are usually quite different from the ones that spur on hunting behavior. See D.L. Hart and R.W. Sussman (2008), *Man the Hunted: Primates, Predators, and Human Evolution* for an extensive review of why predation between species is not the same as, or a good comparison with, aggression between members of the same species.

15. See Douglas Fry (2007), *Beyond War: The Human Potential for Peace* and Walter Goldschmidt (2005), *The Bridge to Humanity: How Affect Hunger Trumps the Selfish Gene* for summaries of cross-cultural examples of the role of experience and cultural context on the expression of aggression.

16. For a review of these studies and more see L.R. Huesmann, E.F. Dubow, and P. Boxer (2011), The transmission of aggressiveness across generations: Biological, contextual, and social learning processes, in P.R. Shaver and M. Mikulincer, eds., *Human Aggression and Violence: Causes, Manifestations, and Consequences*, pp. 123–42.

17. See C.N. DeWall and C.A. Anderson (2011), The general aggression model, in P.R. Shaver and M. Mikulincer eds., *Human Aggression and Violence: Causes, Manifestations, and Consequences*, pp. 15–33.

18. This example is taken from the biologist Robert Sapolsky's 1998 book *Why Zebras Don't Get Ulcers*, which I highly recommend as one of the best, and most accessible, books on hormones and behavior.

19. Azar Gat (2000), The causes and origins of "primitive warfare": Reply to Ferguson, *Anthropological Quarterly* 73(3): 165–68.

20. See chapters 4–20 in C. Campbell, A. Fuentes, K.C. MacKinnon, S. Bearder, and R. Stumpf (2011), *Primates in Perspective*, 2nd ed.

21. R.W. Sussman and P.A. Garber (2011), Cooperation, collective action, and competition in primate social interactions, in C. Campbell, A. Fuentes, K.C. MacKinnon, S. Bearder, and R. Stumpf, eds., *Primates in Perspective,* 2nd ed., pp. 587–98.

22. K. Arnold, O.N. Fraser, and F. Aureli (2011), Postconflict reconciliation, in C. Campbell, A. Fuentes, K.C. MacKinnon, S. Bearder, and R. Stumpf, eds., *Primates in Perspective,* 2nd ed., pp. 608–25.

23. P. Honess and C.M. Marin (2006), Behavioural and physiological aspects of stress and aggression in nonhuman primates, *Neuroscience and Biobehavioral Reviews* 30(3): 390–412.

24. See B.B. Smuts (1987), Gender, aggression and influence, in B.B. Smuts, D.L. Cheney, R.M. Seyfarth, R.W. Wrangham, and T.T. Struhsaker, eds., *Primate Societies,* pp. 400–412.

25. See C. Campbell, A. Fuentes, K.C. MacKinnon, S. Bearder, and R. Stumpf (2011), *Primates in Perspective,* 2nd ed. and M.N. Muller and R.W. Wrangham (2009), *Sexual Coercion in Primates and Humans: An Evolutionary Perspective on Male Aggression against Females* for overviews of this information and for current debates on it. See also B.B. Smuts (1987), Gender, aggression and influence, in B.B. Smuts, D.L. Cheney, R.M. Seyfarth, R.W. Wrangham, and T.T. Struhsaker, eds., *Primate Societies,* pp. 400–412, for an earlier overview on this topic.

26. D.P. Watts, M. Muller, S.J. Amsler, G. Mbabazi, and J.C. Mitani (2006), Lethal intergroup aggression by chimpanzees in Kibale National Park, Uganda, *American Journal of Primatology* 68: 161–80.

27. R.M. Stumpf (2011), Chimpanzees and Bonobos: Inter- and intraspecies diversity, in C. Campbell, A. Fuentes, K.C. MacKinnon, S. Bearder, and R. Stumpf (2011), *Primates in Perspective,* 2nd ed., pp. 340–56.

28. C. Boesch, C. Crockford, I. Herbinger, R. Wittig, I. Moebius, and E. Normand (2008), Intergroup conflicts among chimpanzees in Tai National Park: Lethal violence and the female perspective, *American Journal of Primatology* 70: 519–32.

29. See Rebecca Stumpf's chapter (in C. Campbell, A. Fuentes, K.C. MacKinnon, S. Bearder, and R. Stumpf (2011), *Primates in Perspective,* 2nd ed., pp. 340–56) for a comprehensive overview of chimpanzees and the citations to articles and books where you can find all of the original data for these features of chimpanzee life.

30. See D.L. Hart and R.W. Sussman (2008), *Man the Hunted: Primates, Predators, and Human Evolution* and D. Fry (2007) *Beyond War: The Human Potential for Peace* for extensive discussions of why the chimpanzee aggression data are not particularly useful in reconstructing the role of aggression in human evolution.

31. Interestingly, those who propose this position are largely primatologists working with eastern chimpanzees (many of them living in highly disturbed habitats) as well as some psychologists and political scientists. The majority of anthropologists and biologists who study aggression and warfare do not see this connection as reflecting a common origin and a basal explanation for warfare in humans.

32. C. Boesch, C. Crockford, I. Herbinger, R. Wittig, I. Moebius, and E. Normand (2008), Intergroup conflicts among chimpanzees in Tai National Park: Lethal violence and the female perspective, *American Journal of Primatology* 70: 519–32.

33. See D.L. Hart and R.W. Sussman (2008), *Man the Hunted: Primates, Predators, and Human Evolution* and J. Marks (2002), *What It Means to Be 98% Chimpanzee* for extensive discussion on the problems with using chimpanzees as analogies for human ancestors; and also B.R. Ferguson (2011), Born to live: Challenging killer myths, in R.W. Sussman and R.C. Cloninger, eds., *Origins of altruism and cooperation*, pp. 249–270 (Developments in primatology series, vol. 36, part 2: Progress and prospects).

34. This is of course very debatable in modern warfare where bombs, missiles, unmanned drones, and robots can carry out many of the actual lethal attacks and face-to-face killing is less common. Thus the action of pushing a button or giving a verbal command far away from the battlefield might not require any aggression at all.

35. B. Ferguson (2008b), War before history, in P. deSouza, ed., *The Ancient World at War*.

36. B. Ferguson (2008a), 10 points on war, *Social Analysis* 52(2): 34; B.R. Ferguson (2011), Born to live: Challenging killer myths, in R.W. Sussman and R.C. Cloninger, eds., *Origins of altruism and cooperation*, pp. 249–270 (Developments in primatology series, vol. 36, part 2: Progress and prospects).

37. This includes possible cannibalism, damage to bones from sharp points like a spear, or other types of potential physical violence from the tools and weapons available to our early ancestors. See the following for details: S.C. Anton (2003), Natural History of *Homo erectus, Yearbook of Physical Anthropology* 46: 126–70; W.H. Kimble and L.K. Delezene (2009), "Lucy" redux: A review of research on *Australopithecus afarensis, Yearbook of Physical Anthropology* 52: 2–48; P.L. Walker (2001), A bioarcheological perspective on the history of violence, *Annual Reviews in Anthropology* 30: 573–96; and I. Tattersall (1997), *The Fossil Trail: How We Know What We Think We Know About Human Evolution.* There is also a good deal of contention over how much in the fossil record can be accurately described as cannibalism versus predation (by nonhumans); see D.L. Hart and R.W. Sussman (2008), *Man the Hunted: Primates, Predators, and Human Evolution.*

38. See P.L. Walker (2001), A bioarcheological perspective on the history of violence, *Annual Reviews in Anthropology* 30: 573–96.

39. Azar Gat (2000), The causes and origins of "primitive warfare": Reply to Ferguson, *Anthropological Quarterly* 73(3): 165–68.

40. B. Ferguson (2008a), 10 points on war, *Social Analysis* 52(2): 32–49.

41. Quoted in Dan Jones (2008), Killer instinct, *Nature* 451: 512–15.

42. From P.L. Walker (2001), A bioarcheological perspective on the history of violence, *Annual Reviews in Anthropology* 30: 573–96.

43. Sexual selection of certain traits in males is a common proposal in the animal kingdom. Good examples of male traits assumed to be a result of this are the large antlers of deer and elk and the peacock's flamboyant tail.

44. Because the human species varies so widely in height and weight across populations (see chapter 4) the pattern of males being larger than females generally holds as a species-wide average and within populations, but not necessarily between them.

45. On upper body strength, see W.D. Lassek and S.J.C. Gaulin (2009), Costs and benefits of fat-free muscle mass in men: Relationship to mating success, dietary requirements, and native immunity, *Evolution and Human Behavior* 30: 322–28.

46. The bulk of this basic summary is drawn from the psychologist John Archer's excellent 2004 and 2009 overviews of gender patterns in human aggression because these include meta-analyses of data from several hundred research reports comparing male and female aggression. However the data are largely from Western societies, which might skew some of these perspectives; J. Archer (2009a), Does sexual selection explain human sex differences in aggression? *Behavioral and Brain Sciences* 32: 249–311 and J. Archer (2004), Sex differences in aggression in real world settings: A meta-analytic review, *Review of General Psychology* 8: 291–322.

47. L.R. Huesmann, E.F. Dubow, and P. Boxer (2011), The transmission of aggressiveness across generations: Biological, contextual, and social learning processes, in P.R. Shaver and M. Mikulincer, eds., *Human Aggression and Violence: Causes, Manifestations, and Consequences*, pp. 123–42.

48. J. Archer (2009a), Does sexual selection explain human sex differences in aggression? *Behavioral and Brain Sciences* 32: 249–311.

49. The gender empowerment index is a measure developed by the United Nations to assess women's relative emancipation in different countries. It is calculated from data about the number of women in administrative, professional, managerial, and technical positions, female share of earned income, and female parliamentary/governmental representation.

50. Recent overviews on what is the "norm" for human societies suggest that there is not an evolutionary pattern in aggression. See R.W. Sussman and C.R. Cloninger (2011), *Cooperation and Altruism* and D. Fry (2007), *Beyond War: The Human Potential for Peace*.

51. R.J. Nelson and B.C. Trainor (2007), Neural mechanisms of aggression, *Nature Reviews: Neuroscience* 8: 536–46.

52. B. Ferguson (2008a), 10 points on war, *Social Analysis* 52(2): 34.

53. A. Gibbons (2004), Tracking the evolutionary history of a "warrior" gene, *Science* 304: 818.

54. See the following for a general overview and numerous citations for this information: R.J. Nelson and B.C. Trainor (2007), Neural mechanisms of aggression, *Nature Reviews: Neuroscience* 8: 536–46 and T.F. Denson (2011), A social neuroscience perspective on the neurobiological basis of aggression, in P.R. Shaver and M. Mikulincer, eds., *Human Aggression and Violence: Causes, Manifestations, and Consequences*, pp. 105–20.

55. For the background and original reports on MAOA see the following: A. Gibbons (2004), Tracking the evolutionary history of a "warrior" gene, *Science* 304: 818; T.K. Newman, Y.V. Syagailo, C.S. Barr, J.R. Wendland, M. Champoux, M. Grassle, S.J. Suomi, J.D. Higley, and K. Lesch (2005),

Monoamine oxidase A gene promoter variation and rearing experience influences aggressive behavior in rhesus monkeys, *Biological Psychiatry* 57: 167–72; and N. Alia-Klein, R.Z. Goldstein, A. Kriplani, J. Logan, D. Tomasi, B. Williams, F. Telang, E. Shumay, A. Biegon, I.W. Craig, F. Henn, G. Wang, N.D. Volkow, and J.S. Fowler (2008), Brain monoamine oxidase A activity predicts trait aggression, *Journal of Neuroscience* 28(19): 5099–104.

56. An interesting part of this is that chimpanzees have little to no variation or connection between MAOA expression and aggression, but both rhesus monkeys and humans appear to have something going on here.

57. I (Agustín Fuentes) was genotyped for this gene for a National Geographic special I advised on. It turns out that I have the low-producing alleles; however, I have none of the affects commonly attributed to them. This pattern is common for the majority of individuals with the low-production alleles.

58. For background and additional information on this topic see R.J. Nelson and S. Chiavegatto (2001), Molecular basis of aggression, *TRENDS in Neurosciences* 24(12): 713–19 and R.J. Nelson and B.C. Trainor (2007), Neural mechanisms of aggression, *Nature Reviews: Neuroscience* 8: 536–46.

59. See J. Archer (2006), Testosterone and human aggression: An evaluation of the challenge hypothesis, *Neuroscience and Biobehavioral Reviews* 30: 319–45.

60. See J. Archer (2006), Testosterone and human aggression: An evaluation of the challenge hypothesis, *Neuroscience and Biobehavioral Reviews* 30: 319–45 and L.T. Gettler, S.S. Agustin, and C.W. Kuzawa (2010), Testosterone, physical activity, and somatic outcomes among Filipino males, *American Journal of Physical Anthropology* 142: 590–99.

61. As this is an evolutionary scenario based solely on reproduction, same-sex attraction does not factor in to it. This illustrates some of the serious limitations and possible mistakes inherent in this type of simplistic evolutionary modeling.

62. We tackle and bust this specific myth in greater detail in chapter 6. For the purposes of this chapter we just present it in this simplified form, which has been countered extensively in the biological literature; see M. Borgerhoff-Mulder (2004), Are men and women really so different? *Trends in Ecology, and Evolution* 19(1): 3–6; H. Kokko, and M. Jennions (2003), It takes two to tango. *Trends in Ecology and Evolution* 18(3): 103–104; and Z. Tang-Martinez (2000), Paradigms and primates: Bateman's principle, passive females, and perspectives from other taxa, in S.C. Strum and L.M. Fedigan, eds., *Primate Encounters: Models of Science, Gender, and Society,* pp. 261–74.

63. See G. Arnquist and L. Rowe (2005), *Sexual Conflict.*

64. See M.N. Muller and R.W. Wrangham (2009), *Sexual Coercion in Primates and Humans: An Evolutionary Perspective on Male Aggression against Females.*

65. See C. Campbell (2011), Primate sexuality and reproduction, in C. Campbell, A. Fuentes, K.C. MacKinnon, S. Bearder, and R. Stumpf (2011), *Primates in Perspective,* 2nd ed., pp. 587–98 and C. Boesch, G. Kohou, H. Nene, and L. Vigilant (2006), Male competition and paternity in wild chimpanzees of the Tai Forest, *American Journal of Physical Anthropology* 130: 103–115.

66. S. Suomi (2005), Genetic and environmental factors influencing the expression of impulsive aggression and serotinergic functioning in rhesus monkeys, in R. Tremblay, W.W. Hartup, and J. Archer, eds., *Developmental Origins of Aggression*, pp. 63–82 and B.B. Smuts (1987), Gender, aggression and influence, in B.B. Smuts, D.L. Cheney, R.M. Seyfarth, R.W. Wrangham, and T.T. Struhsaker, eds., *Primate Societies*, pp. 400–412.

67. See chapters 4–20 in C. Campbell, A. Fuentes, K.C. MacKinnon, S. Bearder, and R. Stumpf (2011), *Primates in Perspective*, 2nd ed.

68. D. Sprague (1998), Age, dominance rank, natal status, and tenure among male macaques, *American Journal of Physical Anthropology* 105: 511–21.

69. M. Wilson and R. Daly (2009), Coercive violence by human males against their female partners, in M.N. Muller and R.W. Wrangham, eds., *Sexual Coercion in Primates and Humans: An Evolutionary Perspective on Male Aggression against Females*, pp. 271–91.

70. C. Boehm (2008), Purposive social selection and the evolution of human altruism, *Cross Cultural Research* 42(4): 319–52.

71. Again, D. Fry (2007), *Beyond War: The Human Potential for Peace* is an excellent overview of these data and themes.

72. This subject is the source of extremely contentious debate. I summarize the main points and numbers here but see D. Fry (2007), *Beyond War: The Human Potential for Peace;* N. Chagnon (1998), Life histories, blood revenge, and warfare in a tribal population, *Science* 239: 985–92; N. Chagnon (1990), Reproductive and somatic conflicts of interest in the genesis of violence and warfare among tribesmen, in J. Hass, ed., *The Anthropology of War*, pp. 77–104; and B.R. Ferguson (2011), Born to live: Challenging killer myths, in R.W. Sussman and R.C. Cloninger, eds., *Origins of altruism and cooperation*, pp. 249–270 (Developments in primatology series, vol. 36, part 2: Progress and prospects).

73. S. Beckman, P.I. Erickson, J. Yost, J. Regalado, L. Jaramillo, C. Sparks, M. Iromenga, and K. Long (2009), Life histories, blood revenge, and reproductive success among the Waorani of Ecuador, *PNAS* 106(20): 8134–39.

74. N.J. Enfield and S. Levinson (2006), Introduction: Human Sociality as a New Interdisciplinary Field, in N.J. Enfield and Stephen Levinson, eds., *Roots of Human Sociality: Culture, Cognition, and Interaction.*

75. K. Hill, M. Barton, and M. Hurtado (2009), The emergence of human uniqueness: Characters underlying behavioral modernity, *Evolutionary Anthropology* 18: 187–200.

76. R.C. Marshall (2010), Introduction, in R.C. Marshal, ed., *Cooperation in Social and Economic Life,* pp. vi–xxi.

77. See R.W. Sussman and C.R. Cloninger (2012), *Cooperation and Altruism;* D. Fry (2007), *Beyond War: The Human Potential for Peace;* A. Fuentes (2009b), Re-situating anthropological approaches to the evolution of human behavior, *Anthropology Today* 25(3): 12–17; A. Fuentes (2004), It's not all sex and violence: Integrated anthropology and the role of cooperation and social complexity in human evolution, *American Anthropologist* 106(4): 710–18; R. Oka and A. Fuentes (2010), From reciprocity to trade: How cooperative infrastructures form the basis of human socioeconomic evolution, in R.C. Marshall,

ed., *Cooperation in Social and Economic Life*, pp. 3–28; and M. Nowack and R. Highfield (2011), *Super Cooperators: Evolution, Altruism and Why We Need Each Other to Succeed*.

78. K. Hill, M. Barton, and M. Hurtado (2009), The emergence of human uniqueness: Characters underlying behavioral modernity, *Evolutionary Anthropology* 18: 187–200; A. Fuentes, M. Wyczalkowski, and K.C. MacKinnon (2010), Niche construction through cooperation: A nonlinear dynamics contribution to modeling facets of the evolutionary history in the genus *Homo*, *Current Anthropology* 51(3): 435–44; N.J. Enfield and S. Levinson, eds. (2006), *Roots of Human Sociality: Culture, Cognition, and Interaction*; and J. Henrich, R. Boyd, S. Bowles, C. Camerer, E. Ferh, and H. Gintis, eds. (2004), *Foundations of Human Sociality: Economic Experiments and Ethnographic Evidence from Fifteen Small-Scale Societies*.

79. J. Henrich, R. Boyd, S. Bowles, C. Camerer, E. Ferh, and H. Gintis, eds. (2004), *Foundations of Human Sociality: Economic Experiments and Ethnographic Evidence from Fifteen Small-Scale Societies*.

80. D. Fry (2007), *Beyond War: The Human Potential for Peace*.

81. S. Pinker (2002), *The Blank Slate: The Modern Denial of Human Nature*, p. 317.

82. Big Think April 2010 at http://bigthink.com/ideas/19361.

83. A. Montague (1976), *The Nature of Human Aggression*.

CHAPTER 6

1. See John Gray's Web site for the entire panoply of his endeavors where he offers (for a fee) to help you know who your soul mate is, increase intimacy and passion, create romance with your partner, and enjoy a lifetime of great sex (Marsvenus.com).

2. Marshall Brain (1997), *The Teenager's Guide to the Real World*. You can read this selection online at http://www.bygpub.com/books/tg2rw/chap11excerpt.htm. It is interesting to note that the author cites John Gray (author of *Men Are from Mars, Women Are from Venus*) as a source for this explanation.

3. Nicholas Wade, Pas de deux of sexuality is written in the genes, *New York Times*, April 10, 2007.

4. H. Donnan and F. Magowan (2010), *The Anthropology of Sex*.

5. Janet Shibley Hyde (2005), The gender similarities hypothesis, *American Psychologist* 60(6): 581–92.

6. Lise Eliot (2009) *Pink Brain Blue Brain: How Small Differences Grow into Troublesome Gaps—and What We Can Do about It*.

7. See Anne Fausto-Sterling (2000), *Sexing the Body: Gender Politics and the Construction of Sexuality* for an excellent and extensive overview of these patterns and their explanations and impacts.

8. This is most evident in "drag" shows, where males playing females use hyper-versions of this behavior to mark their femininity.

9. C.M. Pond (1997), The biological origins of adipose tissue in humans, in M.E. Morbeck, A. Galloway , and A.L. Zihlman eds., *The Evolving Female*, pp. 147–62.

10. It is possible that humans are the only species to undergo menopause . . . which is very interesting.

11. Anne Fausto-Sterling (2000), *Sexing the Body: Gender Politics and the Construction of Sexuality.*

12. Again, see Anne Fausto-Sterling (2000), *Sexing the Body: Gender Politics and the Construction of Sexuality* for an excellent overview of these patterns and the medical, and cultural, responses to them.

13. See Robert L. Trivers (1972), Parental investment and sexual selection, in B. Campbell, ed., *Sexual Selection and the Descent of Man 1871–1971*, pp. 136–79.

14. M. Borgerhoff-Mulder and K. Rauch (2009), Sexual conflict in humans: Variations and solutions, *Evolutionary Anthropology* 18: 201–14.

15. Sarah Hrdy (2009), *Mothers and Others: The Evolutionary Origins of Mutual Understanding.*

16. For an extended overview of these ideas and an accurate biological and evolutionary perspective on this topic see H. Kokko and M. Jennions (2003), It takes two to tango, *Trends in Ecology and Evolution* 18(3): 103–104; M. Borgerhoff-Mulder (2004), Are men and women really so different? *Trends in Ecology and Evolution* 19(1): 3–6; Z. Tang-Martinez (2000), Paradigms and primates: Bateman's principle, passive females, and perspectives from other taxa, in S.C. Strum and L.M. Fedigan, eds., *Primate Encounters: Models of Science, Gender, and Society*, pp. 261–74; and M. Borgerhoff-Mulder and K. Rauch (2009), Sexual conflict in humans: Variations and solutions, *Evolutionary Anthropology* 18: 201–14.

17. L. Eliot (2009), *Pink Brain Blue Brain: How Small Differences Grow into Troublesome Gaps—and What We Can Do about It.* See also R.M. Jordan-Young (2011), *Brain Storm: The Flaws in the Science of Sex Differences.*

18. J.L. Wood, D. Heitmiller, N.C. Andreasen, and P. Nopoulos (2008), Morphology of the ventral frontal cortex: relationship to femininity and social cognition, *Cerebral Cortex* 18: 534–40.

19. Remember that on average male brains are slightly larger in size overall, so most of these brain measures first correct for overall volume and then compare the relative sizes of the specific brain areas; otherwise males would basically show up larger in all measures (on average). This correction for relative size is a must for all brain studies, because there is so much variability in human brain size. Healthy human brains can range in size from 1000cc to 2000cc, which is a 100 percent size variance all within the normative range for the species!

20. J.L. Wood, V. Murko, and P. Nopoulos (2008), Ventral frontal cortex in children: Morphology, social cognition and femininity/masculinity, *SCAN* 3: 168–76.

21. Anne Fausto-Sterling (2000), *Sexing the Body: Gender Politics and the Construction of Sexuality.*

22. K. Bishop and D. Wahlsten (1997), Sex differences in the human corpus callosum: Myth or reality? *Neuroscience and Biobehavioral Reviews* 21(5): 581–601.

23. See both Anne Fausto-Sterling (2000), *Sexing the Body: Gender Politics and the Construction of Sexuality,* and R.M. Jordan-Young (2011), *Brain Storm: The Flaws in the Science of Sex Differences.*

24. This pattern can even be found cross-species, especially between humans and dogs. See for example Meg Olmert's 2009 book *Made for Each Other.*

25. For a summary of the oxytocin studies see R. Sanchez, J.C. Parkin, J.Y. Chen, and P.B. Gray (2009), Oxytocin, vasopressin, and human social behavior, in P.T. Ellison and P.B. Gray, eds., *The Endocrinology of Social Relationships,* pp. 319–39.

26. Interestingly, this pattern is also seen in other primates where males do a lot of caretaking of infants.

27. For a summary of the testosterone and prolactin studies in men see P.B. Gray and B.C. Campbell (2009), Human male testosterone, pair bonding and fatherhood, in P.T. Ellison and P.B. Gray, eds., *The Endocrinology of Social Relationships,* pp. 270–93.

28. See C. Wedekind and D. Penn (2000), MHC genes, body odours and odour preferences, *Nephrology Dialysis Transplantation* 15: 1269–71 for a brief overview with references.

29. T.A. Hummer and M.K. McClintock (2009), Putative human pheromone androstadienone attunes the mind specifically to emotional information, *Hormones and Behavior* 55(4): 548–59.

30. W.S.T. Hays (2003), Human pheromones: Have they been demonstrated? *Behavioral Ecology and Sociobiology* 54: 89–97.

31. Lawrence Summers' remarks at the National Bureau of Economic Research Conference on Diversifying the Science and Engineering Workforce, January 14, 2005.

32. Here they used "sex" and "gender" interchangeably. Generally when we refer to sex differences we mean differences between biological males and females, whereas gender differences can sometimes refer only to differences between feminine and masculine social roles. However, the majority of psychological investigations use the terms "gender" and "sex" interchangeably; see E.E. Maccoby and C.N. Jacklin (1974), *The Psychology of Sex Differences.*

33. See Janet Shibley Hyde (2005), The gender similarities hypothesis, *American Psychologist* 60(6): 581–92 for a discussion of this and E.E. Maccoby and C.N. Jacklin (1974), *The Psychology of Sex Differences* for the original reports.

34. Janet Shibley Hyde (2005), The gender similarities hypothesis, *American Psychologist* 60(6): 581–92.

35. This is a well-established procedure for comparing these types of studies.

36. S. Nanda (2000), *Gender Diversity: Cross-Cultural Variations.*

37. W. Wood and A.H. Eagly (2002), A cross-cultural analysis of the behavior of women and men: Implications for the origins of sex differences, *Psychological Bulletin* 128(5): 699–727.

38. Remember, even biological sex is really more of a continuum than an either/or dichotomy.

39. More specifically, everyone from the United States, as gender roles and expectations vary from culture to culture and even across time in the same culture.

40. The anthropologist Serena Nanda tells us that "a cross-cultural perspective makes it clear that there are many different (although probably not unlimited) ways that societies can organize their thinking about sex, gender and sexuality." In this book we are largely focusing on examples of gender in the United States, which may or may not hold true for other parts of the world. There is substantial cross-cultural variation in how gender plays out, with some interesting commonalities and differences. See S. Nanda (2000), *Gender Diversity: Cross-Cultural Variations* for a good overview of gender variants across the planet.

41. W. Wood and A.H. Eagly (2002), A cross-cultural analysis of the behavior of women and men: Implications for the origins of sex differences, *Psychological Bulletin* 128(5): 699–727.

42. Ibid. See also J. Archer (2009a), Does sexual selection explain human sex differences in aggression? *Behavioral and Brain Sciences* 32: 249–311.

43. For background on the fossil and archeological record of human evolution, see A. Fuentes (2009), *Evolution of Human Behavior.*

44. R. Hausmann, L.D. Tyson, and S. Zahidi, eds. (2010), *The Global Gender Gap Report 2010,* World Economic Forum.

45. The World Economic Forum index benchmarks national gender gaps on economic, political, education- and health-based criteria.

46. R. Hausmann, L.D. Tyson, and S. Zahidi, eds. (2010), *The Global Gender Gap Report 2010,* World Economic Forum.

47. See both W. Wood and A.H. Eagly (2002), A cross-cultural analysis of the behavior of women and men: Implications for the origins of sex differences, *Psychological Bulletin* 128(5): 699–727 and J. Archer (2009a), Does sexual selection explain human sex differences in aggression? *Behavioral and Brain Sciences* 32: 249–311 for good overviews and intelligent explanations of this complex scenario.

48. Helen Fisher is a scientific advisor for Chemistry.com; http://chemistry.com/drhelenfisher/InterviewDrFisher.aspx#2.

49. Although there are also many who would argue that the female is more drawn to this bonded state and the male resists it. See for example the classic D. Symons (1979), *The Evolution of Human Sexuality* and D.M. Buss and D.P. Schmitt (1993), Sexual strategies theory: A contextual evolutionary analysis of human mating, *Psychological Review* 100: 204–32.

50. These facets of relationships and sex are covered in fascinating detail in C. Ryan and C. Jetha (2010), *Sex at Dawn: The Prehistoric Origins of Modern Sexuality.*

51. Walter Goldschmidt (2005), *The Bridge to Humanity: How Affect Hunger Trumps the Selfish Gene.*

52. This notion is increasingly popular in studies of human evolution. For an overview of these ideas see A. Fuentes (2009a), *Evolution of Human Behavior.*

53. Biologically speaking, that is. Of course there are various different psychological and social elements involved as well as religious ones.

54. For recent versions of this approach see B. Chapais (2008), *Primeval Kinship: How Pair-Bonding Gave Birth to Human Society* and O. Lovejoy

(2009), Reexamining human origins in light of *Ardipithecus ramidus, Science* 326: 108–15.

55. See A. Fuentes (1999), Re-evaluating primate monogamy, *American Anthropologist* 100(4): 890–907 and A. Fuentes (2002), Patterns and trends in primate pair bonds, *International Journal of Primatology* 23(4): 953–78.

56. See multiple chapters in P.T. Ellison and P.B. Gray, eds. (2009) *The Endocrinology of Social Relationships,* pp. 270–93 and J.T. Curtis and Z. Wang (2003), The neurochemistry of pair bonding, *Current Directions in Psychological Science* 12(2): 49–53.

57. See multiple chapters in P.T. Ellison and P.B. Gray, eds. (2009), *The Endocrinology of Social Relationships,* pp. 270–93 and J.T. Curtis and Z. Wang (2003), The neurochemistry of pair bonding, *Current Directions in Psychological Science* 12(2): 49–53.

58. However, just being a relative does not automatically generate a pair bond.

59. B. Chapais (2008), *Primeval Kinship: How Pair-bonding Gave Birth to Human Society.*

60. A. Fuentes (2009), *Evolution of Human Behavior* and A. Fuentes, M. Wyczalkowski, and K.C. MacKinnon (2010), Niche construction through cooperation: A nonlinear dynamics contribution to modeling facets of the evolutionary history in the genus *Homo, Current Anthropology* 51(3): 435–44.

61. D.P. Schmitt (2005), Sociosexuality from Argentina to Zimbabwe: A 48-nation study of sex, culture, and strategies of human mating, *Behavioral and Brain Sciences* 28: 247–311.

62. D.P. Barash and J.E. Lipton (2001), *The Myth of Monogamy.*

63. For overviews of these data see D.P. Barash and J.E. Lipton (2001), *The Myth of Monogamy;* D.P. Schmitt (2005), Sociosexuality from Argentina to Zimbabwe: A 48-nation study of sex, culture, and strategies of human mating, *Behavioral and Brain Sciences* 28: 247–311; H. Fisher (1992), *Anatomy of Love: The Natural History of Monogamy, Adultery, and Divorce;* and A. Fuentes (1999), Re-evaluating primate monogamy, *American Anthropologist* 100 (4): 890–907.

64. U.H. Reichard and C. Boesch, eds. (2003), *Monogamy: Mating Strategies and Partnerships in Birds, Humans and Other Mammals.*

65. For a well-written recent overview see S. Squire (2008), *I Don't: A Contrarian History of Marriage.*

66. Early Judaism, early Christianity, some modern Christian sects, and modern-day Islam all sanctioned polygynous (one male with multiple wives) as well as monogamous marriage. In all instances females always have be monogamous to their husbands and in all cases husbands are supposed to be exclusive with their wife (or wives in polygynous marriages).

67. See D.P. Barash and J.E. Lipton (2001), *The Myth of Monogamy;* H. Fisher (1992), *Anatomy of Love: The Natural History of Monogamy, Adultery, and Divorce;* and S. Squire (2008), *I Don't: A Contrarian History of Marriage.*

68. See also C. Ryan and C. Jetha (2010), *Sex at Dawn: The Prehistoric Origins of Modern Sexuality.*

69. Anne Fausto-Sterling (2000), *Sexing the Body: Gender Politics and the Construction of Sexuality.*

70. What men and women are sexually includes how they see themselves sexually, who they are attracted to, what they do sexually, and so on.

71. H. Donnan and F. Magowan (2010), *The Anthropology of Sex.*

72. D. Herbenick, M. Reece, V. Schick, S.A. Sanders, B. Dodge, and J.D. Fortenberry (2010), Sexual behavior in the United States: Results from a national probability sample of men and women ages 14–94, *Journal of Sexual Medicine* 7(suppl. 5): 255–65.

73. For a more in-depth description of this (and most ideas covered in this section) see Janell L. Carroll's excellent 2009 human sexuality textbook (*Sexuality Now: Embracing Diversity*, 3rd ed.) and also J. DeLamater and W.N. Friedrich (2002), Human sexual development, *Journal of Sex Research* 39(1): 10–14.

74. Menarche is the onset of menstrual cycling in females.

75. S.W. Gangestad and G.J. Scheyd (2005), The evolution of human physical attractiveness, *Annual Review of Anthropology* 34: 523–48.

76. Facial features and symmetry are a popular focus, but recently, it is less than clear if this hypothesis is really supported by the broader scientific literature. See Jonah Lehrer's December 13, 2010 *New Yorker* magazine essay, "The truth wears off"; http://www.newyorker.com/reporting/2010/12/13/101213fa_fact_lehrer?currentPage-all.

77. See S.W. Gangestad and G.J. Scheyd (2005), The evolution of human physical attractiveness, *Annual Review of Anthropology* 34: 523–48 for a review of positive accounts of waist-to-hip ratios; see also G.H. Shepard and D.W. Yu (1998), Is beauty in the eye of the beholder? *Nature* 326: 321–22 and J. Freese and S. Meland (2002), Seven tenths incorrect: Heterogeneity and change in the waist-to-hip ratios of *Playboy* centerfold models and Miss America pageant winners, *The Journal of Research* 39(2): 133–38.

78. For an expansion and real assessment of this see J. Henrich, S.J. Heine, and A. Norenzayan (2010), The weirdest people in the world? *Behavioral and Brain Sciences* 33: 61–135.

79. See C. Wedekind and D. Penn (2000), MHC genes, body odours and odour preferences, *Nephrology Dialysis Transplantation* 15: 1269–71 for a brief overview with references.

80. M.J. Elders (2010), Sex for health and pleasure throughout the lifetime, *Journal of Sexual Medicine* 7(suppl. 5): 248.

81. See A.C. Kinsey, W.B. Pomeroy, and C.E. Martin (1948), *Sexual Behavior in the Human Male*, and A.C. Kinsey, W.B. Pomeroy, C.E. Martin, and P.H. Gebhard (1953), *Sexual Behavior in the Human Female*. The movie about Kinsey's life and work *(Kinsey)* is also well worth seeing.

82. The 2010 National Survey of Sexual Health and Behavior is a US study only. There are very, very few cross cultural studies that look at sexual behavior in this level of detail.

83. The statistics are summarized data presented in the categories used by the main data article: D. Herbenick, M. Reece, V. Schick, S.A. Sanders, B. Dodge, and J.D. Fortenberry (2010), Sexual behavior in the United States: Results from

a national probability sample of men and women ages 14–94, *Journal of Sexual Medicine* 7(suppl. 5): 255–65.

84. The authors of the study make the point that their survey undersampled homosexually oriented populations.

85. See S.T. Lindau, L.P. Schumm, E.O. Laumann, W. Levinson, C.A. O'Muircheartaigh, and L.J. Waite (2007), A study of sexuality and health among older adults in the United States, *New England Journal of Medicine* 357(8): 762–74 and A. Brewis and M. Meyer (2005), Marital coitus across the life course, *Journal of Biocultural Science* 37: 499–518.

86. D.P. Schmitt (2005), Sociosexuality from Argentina to Zimbabwe: A 48-nation study of sex, culture, and strategies of human mating, *Behavioral and Brain Sciences* 28: 247–311.

87. In fact, there is good evidence that gender roles and cultural pressures affect the way people respond to questions about sexuality. See M.G. Alexander and T.D. Fisher (2002), Truth and consequences: Using the bogus pipeline to examine sex differences in self-reported sexuality, *Journal of Sex Research* 40(1): 27–35.

88. David P. Schmitt (2005), Sociosexuality from Argentina to Zimbabwe: A 48-nation study of sex, culture, and strategies of human mating, *Behavioral and Brain Sciences* 28: 247–311.

89. See D.M. Buss and D.P. Schmitt (1993), Sexual strategies theory: A contextual evolutionary analysis of human mating, *Psychological Review* 100: 204–32 and W.C. Pedersen, L.C. Miller, A.D. Putcha-Bhagavatula, and Y. Yang (2002), Evolved sex differences in the number of partners desired? The long and the short of it, *Psychological Science* 13(2): 157–61.

90. See J. Henrich, S.J. Heine, and A. Norenzayan (2010), The weirdest people in the world? *Behavioral and Brain Sciences* 33: 61–135.

91. See H. Donnan and F. Magowan (2010), *The Anthropology of Sex* and Janell L. Carroll (2009), *Sexuality Now: Embracing Diversity*, 3rd ed.

92. D.P. Schmitt (2005), Sociosexuality from Argentina to Zimbabwe: A 48-nation study of sex, culture, and strategies of human mating, *Behavioral and Brain Sciences* 28: 247–311.

93. Although it is worth noting that some ape species, especially the bonobos (*Pan paniscus*) do engage in very high levels of sexual activity (as noted in chapter 5). This suggests the possibility that sexuality might be more complex in other animals as well, but as we have seen, sex is very difficult to study.

94. See C. Ryan and C. Jetha (2010), *Sex at Dawn: The Prehistoric Origins of Modern Sexuality* and Z. Tang-Martinez (2000), Paradigms and primates: Bateman's principle, passive females, and perspectives from other taxa, in S.C. Strum and L.M. Fedigan, eds., *Primate Encounters: Models of Science, Gender, and Society*, pp. 261–74.

95. R.M. Jordan-Young (2011), *Brain Storm: The Flaws in the Science of Sex Differences*.

96. The baseball throw from the shoulder is a very difficult thing for the human body; it is not an innate movement and you have to be well trained to do it effectively. On a first exposure to throwing a small ball most humans

will just throw from the elbow (called "throwing like a girl") rather than pull the arm back and use the shoulder rotation to gain speed and power for the throw.

97. See Tracy Clark-Flory's interesting series of essays on this at http://www .salon.com/life/sex/index.html?story = /mwt/feature/2011/08/06/monogamy.

CHAPTER 7

1. See http://dictionary.reference.com/browse/lie for a basic definition, or better yet pull a dictionary off the shelf and see how the entry reads (they vary pretty widely).

2. A.D. Jennions and A.P. Møllers (2002), Relationships fade with time: A meta-analysis of temporal trends in publication in ecology and evolution, *Proceedings of the Royal Society B* 269: 43–48.

3. See for example, Jonah Lehrer's December 13, 2010, *New Yorker* magazine essay, "The truth wears off"; http://www.newyorker.com/reporting/2010/ 12/13/101213fa_fact_lehrer?currentPage=all.

4. See Jon Mark's excellent 2009 book *Why I Am Not a Scientist* for an in-depth discussion of examples over the last century of scientists intentionally pushing ideas that reinforce these myths even when their own data refuted them.

5. Unless there is an astronaut or pilot reading this.

6. http://www.salon.com/life/this_week_in_blackness/2011/01/04/huck_finn_ n_word.

7. For a recent overview, see C. Hertzman and T. Boyce (2010), How experience gets under the skin to create gradients in developmental health, *Annual Review of Public Health* 31: 329–47.

APPENDIX

1. Major scientific journals like *Science* and *Nature* can now be searched online as far back as the 1880s—you can pull up digitized copies of the core scientific publications from the last 130 years!

2. http://onlinebooks.library.upenn.edu, hosted by the University of Pennsylvania.

3. http://darwin-online.org.uk.

4. http://www.doaj.org. As of January 2011 . . . this number will be much larger in 2012, 2013 and beyond.

5. For good examples see http://www.cdc.gov, http://physanth.org, http:// aaanet.org, http://www.apa.org, and http://www.aaas.org.

6. For a great anthropology and science blog, see http://blogs.plos.org/ neuroanthropology.

7. Peer review is a process in the academic world whereby manuscripts are reviewed and critiqued before they are published or rejected. The most common pattern is that a manuscript is submitted, reviewed, and extensively revised in light of those reviews before it is ever published. This process (ideally) makes the information in the article and the conclusions drawn more rigorous than when no peer review is conducted.

8. Encyclopedias and popular books produced by university presses (such as the University of California Press, which is publishing this book) are excellent sources because the book manuscripts are more rigorously reviewed by experts in the appropriate fields than are the manuscripts for books published by purely popular presses.

9. For free lectures by public intellectuals and artists, see http://www.ted.com.

Bibliography

Achter, P., and Condit, C. (2000, May—June). Not so black and white. *American Scientist*. Retrieved from http://www.americanscientist.org/bookshelf/pub/not-so-black-and-white.

Adovasio, J.M., Olga, S., and Jake, P. (2007). *The invisible sex: Uncovering the true roles of women in prehistory*. New York: Smithsonian Books.

Alexander, M.G., and Fisher, T.D. (2002). Truth and consequences: Using the bogus pipeline to examine sex differences in self-reported sexuality. *Journal of Sex Research* 40(1): 27–35.

Alia-Klein, N., Goldstein, R.Z., Kriplani, A., Logan, J., Tomasi, D., Williams, B., Telang, F., Shumay, E., Biegon, A., Craig, I.W., Henn, F., Wang, G., Volkow, N.D., and Fowler, J.S. (2008). Brain monoamine oxidase A activity predicts trait aggression. *Journal of Neuroscience* 28(19): 5099–5104.

American Anthropological Association. (1998). AAA statement on race. Retrieved from http://www.aaanet.org/issues/policy-advocacy/AAA-Statement-on-Race.cfm.

Angier, N. (2000, October 10). A conversation with Paul Ehrlich: On human nature and the evolution of culture. *New York Times*. Retrieved from http://www.nytimes.com/2000/10/10/science/a-conversation-with-paul-ehrlich-on-human-nature-and-the-evolution-of-culture.html.

Anton, S.C. (2003). Natural history of *Homo erectus*. *Yearbook of Physical Anthropology* 46: 126–170.

Aquinas, T. (1873). *Summa theologiae*. Rome: Leonine.

Archer, J. (2004). Sex differences in aggression in real world settings: A meta-analytic review. *Review of General Psychology* 8: 291–322.

———. (2006). Testosterone and human aggression: An evaluation of the challenge hypothesis. *Neuroscience and Biobehavioral Reviews* 30: 319–345.

———. (2009a). Does sexual selection explain human sex differences in aggression? *Behavioral and Brain Sciences* 32: 249–311.

———. (2009b). The nature of human aggression. *International Journal of Law and Psychiatry* 32: 202–208.

Armelagos, G.J. (2005). The slavery hypothesis—natural selection and scientific investigation: A commentary. *Transforming Anthropology* 30(2): 119–124.

Arnold, K., Fraser, O.N., and Aureli, F. (2011). Postconflict reconciliation. In C. Campbell, A. Fuentes, K.C. MacKinnon, S. Bearder, and R. Stumpf, eds., *Primates in perspective*, 2nd ed., pp. 587–598. New York: Oxford University Press.

Arnquist, G., and Rowe, L. (2005). *Sexual conflict.* Princeton, NJ: Princeton University Press.

Bach, P.B., Hoangmai, H.P., Schrag, D., Tate, R.C., and Hargraves, J.L. (2004). Primary care physicians who treat blacks and whites. *New England Journal of Medicine* 351: 6.

Barash, D.P., and Lipton, J.E. (2001). *The myth of monogamy.* New York: W.H. Freeman.

Basic information about HIV and AIDS. (2010). Retrieved from http://www.cdc.gov/hiv/topics/basic/index.htm.

Beckman, S., Erickson, P.I., Yost, J., Regalado, J., Jaramillo, L., Sparks, C., Iromenga, M., and Long, K. (2009). Life histories, blood revenge, and reproductive success among the Waorani of Ecuador. *PNAS* 106(20): 8134–8139.

Bernstein, I.S. (2007). Social mechanisms in the control of primate aggression. In C.J. Campbell, A. Fuentes, K.C. MacKinnon, M. Panger, and S.K. Bearder, eds., *Primates in Perspective*, pp. 562–570. New York: Oxford University Press.

Bishop, K., and Wahlsten, D. (1997). Sex differences in the human corpus callosum: Myth or reality? *Neuroscience and Biobehavioral Reviews* 21(5): 581–601.

Boas, F. (1930). Anthropology. *Encyclopedia of the social sciences,* vol. 2. Edited by E.A. Seligman and A. Johnson, pp. 73–110. New York: Macmillan.

Boehm, C. (2008). Purposive social selection and the evolution of human altruism. *Cross Cultural Research* 42(4): 319–352.

Boesch, C., Crockford, C., Herbinger, I., Wittig, R., Moebius, I., and Normand, E. (2008). Intergroup conflicts among chimpanzees in Tai National Park: Lethal violent and the female perspective. *American Journal of Primatology* 70: 519–532.

Boesch, C., Kohou, G., Nene, H., and Vigilant, L. (2006). Male competition and paternity in wild chimpanzees of the Tai Forest. *American Journal of Physical Anthropology* 130: 103–115.

Borgerhoff-Mulder, M. (2004). Are men and women really so different? *Trends in Ecology and Evolution* 19(1): 3–6.

Borgerhoff-Mulder, M., and Rauch, K. (2009). Sexual conflict in humans: Variations and solutions. *Evolutionary Anthropology* 18: 201–214.

Boyd, C.A. (2004). Was Thomas Aquinas a sociobiologist? Thomistic natural law, rational goods, and sociobiology. *Zygon* 39(3): 659–680.

Brain, M. (1997). *The Teenager's Guide to the Real World*. Raleigh, NC: BYG Publishing.

Brewis, A. and Meyer, M. (2005). Martial coitus across the life course. *Journal of Biocultural Science* 39: 499–518.

Buss, D.M., and Schmitt, D.P. (1993). Sexual strategies theory: A contextual evolutionary analysis of human mating. *Psychological Review* 100: 204–232.

Campbell, C. (2011). Primate sexuality and reproduction. In C. Campbell, A. Fuentes, K.C. MacKinnon, S. Bearder, and R. Stumpf, eds., *Primates in perspective*, 2nd ed., pp. 587–598. New York: Oxford University Press.

Campbell, C., Fuentes, A., MacKinnon, K.C., Bearder, S., and Stumpf, R. (2011). *Primates in perspective*, 2nd ed. New York: Oxford University Press.

Carroll, J.L. (2010). *Sexuality now: Embracing diversity*, vol. 3. Belmont, CA: Wadsworth Cengage Learning.

Chagnon, N. (1990). Reproductive and somatic conflicts of interest in the genesis of violence and warfare among tribesmen. In J. Hass, ed., *The anthropology of war*, pp. 77–104. Cambridge: Cambridge University Press.

———. (1998). Life histories, blood revenge, and warfare in a tribal population. *Science* 239: 985–992.

Chapais, B. (2008). *Primeval kinship: How pair-bonding gave birth to human society*. Cambridge, MA: Harvard University Press.

Chomsky, N., and Foucault, M. (2006). *The Chomsky-Foucault debate: On human nature*. New York: New Press.

Comfort, N.C. (2001, June). Are genes real? *Natural History* 110(5): 28–38.

Coppola, F.F., and Mutrix, G. (producers), and Condon, B. (screenwriter/director). (2004). *Kinsey* [Motion Picture]. United States: Fox Searchlight Pictures.

Curtis, J.T., and Wang, Z. (2003). The neurochemistry of pair bonding. *Current Directions in Psychological Science* 12(2): 49–53.

Darwin, C. (1859). *On the origin of species*. New York: New American.

Darwin, C., and Wallace, A.R. (1858). On the tendency of species to form varieties; And on the perpetuation of varieties and species by natural means of selection. *The Linnean Society of London*. Retrieved from http://wallacefund.info/the-1858-darwin-wallace-paper.

Dawkins, C.R. (1976). *The selfish gene*. Oxford: Oxford University Press.

———. (1982). *The extended phenotype*. Oxford: Oxford University Press.

DeLamater, J., and Friedrich, W.N. (2002). Human sexual development. *Journal of Sex Research* 39(1): 10–14.

Denson, T.F. (2011). A social neuroscience perspective on the neurobiological basis of aggression. In P.R. Shaver and M. Mikulincer, eds., *Human Aggression and Violence: Causes, Manifestations, and Consequences*, pp. 105–120. Washington, DC: American Psychological Association.

DeWall, C.N., and Anderson, C.A. (2011). The general aggression model. In P.R. Shaver and M. Mikulincer, eds., *Human aggression and violence: Causes, manifestations, and consequences*, pp. 15–33. Washington, DC: American Psychological Association.

Dobzhansky, T. (1972). On the evolutionary uniqueness of man. In T. Dobzhansky, M.K. Hect, and W.C. Steere, eds., *Evolutionary biology*, vol. 6, pp. 415–430. New York: Appleton-Century-Crofts.

Donnan, H., and Magowan, F. (2010). *The anthropology of sex*. Oxford: Berg Publishers.

Dressler, W.W., Oths, K.S., and Gravlee, C.C. (2005). Race and ethnicity in public health research: Models to explain health disparities. *Annual Review of Anthropology* 34: 231–252.

Dufour, D.L. (2006). The 23rd annual Raymond Pearl memorial lecture: Biocultural approaches in human biology. *American Journal of Human Biology* 18(1): 1–9.

Duster, T. (2005). Race and reification in science. *Science* 307: 1050–1051.

Edgar, H.J.H., and Hunley, K.L. (2009). Race reconciled? How biological anthropologists view human variation. *American Journal of Physical Anthropology* 139: 1–4.

Ehrlich, P. (1999). *Human natures: Genes, cultures and the human prospect*. Washington, DC: Island Press.

Ehrlich, P., and Feldman, M. (2002). Genes and culture: What creates our behavioral phenomena? *Current Anthropology* 44(1): 87–107.

Elders, M.J. (2010). Sex for health and pleasure through the lifetime. *Journal of Sexual Medicine* 7(suppl. 5): 248.

Eliot, L. (2009). *Pink brain, blue brain: How small differences grow into troublesome gaps—and what we can do about it*. New York: Houghton Mifflin Harcourt.

Ellison, P.T., and Gray, P.B., eds. (2009). *The endocrinology of social relationships*. Cambridge, MA: Harvard University Press.

Enfield, N.J., and Levinson, S., eds. (2006). *Roots of human sociality: Culture, cognition, and interaction*. Oxford: Berg Publishers.

Entine, J. (2000). *Taboo: Why black athletes dominate sports and why we're afraid to talk about it*. New York: Public Affairs Books.

———. (2007). *Abraham's children: Race, identity, and the DNA of the chosen people*. New York: Grand Central Publishing.

Fausto-Sterling, A. (2000). *Sexing the body: Gender politics and the construction of sexuality*. New York: Basic Books.

Feresin, E. (2009, October 30). Lighter sentence for murderer with "bad genes." *Scientific American*. Retrieved from http://www.scientificamerican.com/article.cfm?id=lighter-sentence-for-murderer-

Ferguson, B. (2008a). 10 points on war. *Social Analysis* 52(2): 32–49.

———. (2008b). War before history. In P. deSouza, ed., *The ancient world at war*, pp. 15–28. New York: Thames and Hudson.

Ferguson, B.R. (2011). Born to live: Challenging killer myths, in R.W. Sussman and R.C. Cloninger, eds., *Origins of altruism and cooperation*, pp. 249–270 (Developments in Primatology series, vol. 35, part 2: Progress and prospects). New York: Springer Press.

Fisher, H. (1992). *Anatomy of love: The natural history of monogamy, adultery, and divorce*. New York: W.W. Norton.

————. (n.d.). Interview. Retrieved from http://chemistry.com/drhelenfisher/InterviewDrFisher.aspx#2.

Flack, J.C., Girvan, M., de Waal, F.B.M., and Krakauer, D.C. (2006). Policing stabilizes construction of social niches in primates. *Nature* 439: 426–429.

Fox-Keller, E. (2000). *The century of the gene.* Cambridge, MA: Harvard University Press.

Freese, J., and Meland, S. (2002). Seven tenths incorrect: Heterogeneity and change in the waist-to-hip ratios of *Playboy* centerfold models and Miss America pageant winners. *The Journal of Research* 39(2): 133–138.

Frothingham, O.B. (1900, June). Real case of the remonstrants against woman suffrage. *Arena* 2: 175–181.

Fry, D. (2007). *Beyond war: The human potential for peace.* New York: Oxford University Press.

Fuentes, A. (1999), Re-evaluating primate monogamy. *American Anthropologist* 100(4): 890–907.

————. (2002). Patterns and trends in primate pair bonds. *International Journal of Primatology* 23(4): 953–978.

————. (2004). It's not all sex and violence: Integrated anthropology and the role of cooperation and social complexity in human evolution. *American Anthropologist* 106(4): 710–718.

————. (2008). *Core concepts in biological anthropology.* New York: McGraw-Hill.

————. (2009a). *Evolution of human behavior.* New York: Oxford University Press.

————. (2009b). Re-situating anthropological approaches to the evolution of human behavior. *Anthropology Today* 25(3): 12–17.

————. (2011). *Biological anthropology: Concepts and connections,* 2nd ed. New York: McGraw-Hill.

Fuentes, A., Wyczalkowski, M., and MacKinnon, K.C. (2010). Niche construction through cooperation: A nonlinear dynamics contribution to modeling facets of the evolutionary history in the genus *Homo. Current Anthropology* 51(3): 435–444.

Futuyma, D. (2009). *Evolution,* 2nd ed. Sunderland, MA: Sinauer Associates.

Gangestad, S.W., and Scheyd, G.J. (2005). The evolution of human physical attractiveness. *Annual Review of Anthropology* 34: 523–548.

Gat, A. (2000). The causes and origins of "primitive warfare": Reply to Ferguson. *Anthropological Quarterly* 73(3): 165–168.

Geertz, C. (1973). *The interpretation of cultures.* New York: Basic Books.

————. (1983). *Local knowledge: Further essays in interpretive anthropology.* New York: Basic Books.

Gettler, L.T., Agustin, S.S., and Kuzawa, C.W. (2010). Testosterone, physical activity, and somatic outcomes among Filipino males. *American Journal of Physical Anthropology* 142: 590–599.

Gibbons, A. (2004). Tracking the evolutionary history of a "warrior" gene. *Science* 304: 818.

Goldschmidt, W. (2005). *The bridge to humanity: How affect hunger trumps the selfish gene.* New York: Oxford University Press.

Goodman, A.H., and Leatherman, T.L., eds. (1998). *Building a new biocultural synthesis.* Ann Arbor: University of Michigan Press.

Gould, S.J. (2002). *The structure of evolutionary theory.* Cambridge, MA: Belknap Press.

Gravlee, C.C. (2009). How race becomes biology: Embodiment of social inequality. *American Journal of Physical Anthropology* 139: 47–57.

Gravlee, C.C., Non, A.L., and Mulligan, C.J. (2009). Genetic ancestry, social classification, and racial inequalities in blood pressure in southeastern Puerto Rico. *PLoS ONE* 4(9): e6821.

Gray, P.B., and Campbell, B.C. (2009). Human male testosterone, pair bonding, and fatherhood. In P.T. Ellison and P.B. Gray, eds., *The endocrinology of social relationships,* pp. 270–293. Cambridge, MA: Harvard University Press.

Gray, T. (1747). Ode on a distant prospect of Eton College. Retrieved from http://www.thomasgray.org/cgi-bin/display.cgi?text = odec.

Hagen, E. (1996). AAPA statement on the biological aspects of race. *American Journal of Physical Anthropology* 101: 569–570. Retrieved from http://physanth.org/association/position-statements/biological-aspects-of-race/?searchterm=race.

Harrison, G. (2010). *Race and reality: What everyone should know about our biological diversity.* New York: Prometheus Books.

Hart, D.L. and Sussman, R.W. (2008) Man the hunted: Primates, predators, and human evolution. New York: Basic Books.

Hartigan Jr., J. (2010). *Race in the 21st century: Ethnographic approaches.* New York: Oxford University Press.

Hausmann, R., Tyson, L.D., and Zahidi, S., eds. (2010). *The global gender gap report 2010.* World Economic Forum. Retrieved from www3.weforum.org/docs/WEF_GenderGap_Report_2010.pdf.

Hays, W.S.T. (2003). Human pheromones: Have they been demonstrated? *Behavioral Ecology and Sociobiology* 54: 89–97.

Henrich, J., Boyd, R., Bowles, S., Camerer, C., Ferh, E., and Gintis, H., eds. (2004). *Foundations of human sociality: Economic experiments and ethnographic evidence from fifteen small-scale societies.* New York: Oxford University Press.

Henrich, J., Heine, S.., and Norenzayan, A. (2010). The weirdest people in the world? *Behavioral and Brain Sciences* 33: 61–135.

Herbenick, D., Reece, M., Schick, V., Sanders, S.A., Dodge, B., and Fortenberry, J.D. (2010). Sexual behavior in the United States: Results from a national probability sample of men and women ages 14–94. *Journal of Sexual Medicine* 7(suppl. 5): 255–265.

Herrnstein, R., and Murray, C. (1994). *The bell curve: Intelligence and class structure in American life.* New York: Free Press.

Hertzman, C., and Boyce, T. (2010). How experience gets under the skin to create gradients in developmental health. *Annual Review of Public Health* 31: 329–347.

Hill, K., Barton, M., and Hurtado, M. (2009). The emergence of human uniqueness: Characters underlying behavioral modernity. *Evolutionary Anthropology* 18: 187–200.

Hobbes, T. (1651). *Leviathan.* London: Andrew Crooke.

Honess, P., and Marin, C.M. (2006). Behavioural and physiological aspects of stress and aggression in nonhuman primates. *Neuroscience and Biobehavioral Reviews* 30(3): 390–412.

Hrdy, S.B. (2009). *Mother and others: The evolutionary origins of mutual understanding.* Cambridge, MA: Harvard University Press.

Hruschka, D.J., Lende, D., and Worthman, C.M. (2005). Biocultural dialogues: Biology and culture in psychological anthropology. *Ethos* 33(1): 1–19.

Huesmann, L.R., Dubow, E.F., and Boxer, P. (2011). The transmission of aggressiveness across generations: Biological, contextual, and social learning processes. In P.R. Shaver and M. Mikulincer, eds., *Human aggression and violence: Causes, manifestations, and consequences,* pp. 123–142. Washington, DC: American Psychological Association.

Hummer, T.A., and McClintock, M.K. (2009). Putative human pheromone androstadienone attunes the mind specifically to emotional information. *Hormones and Behavior* 55(4): 548–559.

Hyde, J.S. (2005). The gender similarities hypothesis. *American Psychologist* 60(6): 581–592.

Ingold, T. (1996). Situating action v: The history and evolution of bodily skills. *Ecological Psychology* 8(2): 171–182.

———. (2000). *The perception of the environment: Essays on livelihood, dwelling and skill.* London: Routledge.

Jablonka, E., and Lamb, M. (2005). *Evolution in four dimensions: Genetic, epigenetic, behavioral, and symbolic variation in the history of life.* Cambridge: MIT Press.

Jablonski, N.G. (2004). The evolution of human skin and skin color. *Annual Reviews in Anthropology* 33: 585–623.

———. (2006). *Skin: A natural history.* Berkeley: University of California Press.

Jablonski, N.G., and Chaplin, G. (2000). The evolution of human skin color. *Journal of Human Evolution* 39: 57–106.

Jackson, P.L., Meltzoff, A.N., and Decety, J. (2006). Neural circuits involved in imitation and perspective-taking. *NeuroImage* 31(1): 429–439.

Jantz, R.L., and Meadows Jantz, L. (2002). Secular change in craniofacial morphology. *American Journal of Human Biology* 12: 327–338.

Jennions, A.D., and Møllers, A.P. (2002). Relationships fade with time: A meta-analysis of temporal trends in publication in ecology and evolution. *Proceedings of the Royal Society B* 269: 43–48.

Jones, D. (2008). Killer instinct. *Nature* 451: 512–515.

Jordan-Young, R.M. (2011). *Brain storm: The flaws in the science of sex differences.* Cambridge, MA: Harvard University Press.

Kahn, J. (2007, August). Race in a bottle. *Scientific American* 297(2): 40–45.

Kaufman, J. (2006). The anatomy of a medical myth. Retrieved from http://raceandgenomics.ssrc.org/Kaufman.

Kimble, W.H., and Delezne, L.K. (2009). "Lucy" redux: A review of research on *Australopithecus afarensis. Yearbook of Physical Anthropology* 52: 2–48.

Kinsey, A.C., Pomeroy, W.B., and Martin, C.E. (1948). *Sexual behavior in the human male.* Philadelphia: W.B. Saunders.

Kinsey, A.C., Pomeroy, W.B., Martin, C.E., and Gebhard, P.H. (1953). *Sexual behavior in the human female.* Philadelphia: W.B. Saunders.

Kittles, R.A., and Weiss, K.M. (2003). Race, genes and ancestry: Implications for defining disease risk. *Annual Reviews in Human Genetics* 4: 33–67.

Kokko, H., and Jennions, M. (2003). It takes two to tango. *Trends in Ecology and Evolution* 18(3): 103–104.

Kottack, C.P., and Kozaitis, K.A. (2003). *On being different: Diversity and multiculturalism in the North American mainstream.* New York: McGraw-Hill.

Krieger, N., and Sidney, S. (1996). Racial discrimination and blood pressure: The CARDIA study of young black and white adults. *American Journal of Public Health* 86(10): 1370–1378.

Krieger, N., Smith, K., Naishadham, D., Hartman, C., and Barbeau, E.M. (2005). Experiences of discrimination: Validity and reliability of a self-report measure for population health research on racism and health. *Social Science and Medicine* 61: 1576–1596.

Kroeber, A., and Kluckhohn, C. (1952). Culture: A critical review of concepts and definitions. *Papers of the Peabody Museum, Harvard University* 47: 357.

Laland, K., Kendall, J., and Brown, G. (2007). The niche construction perspective: Implications for evolution and human behavior. *Evolutionary Psychology* 5: 51–66.

Lapchick, R., Kaiser, C., Caudy, D., and Wang, W. (2010). The 2010 racial and gender report card: Major League Baseball. Retrieved from http://www.tidesport.org/RGRC/2010/2010_MLB_RGRC_updated.pdf.

Lapchick, R., Kaiser, C., Russell, C., and Welch, N. (2010). The 2010 racial and gender report card: National Basketball Association. Retrieved from http://www.tidesport.org/RGRC/2010/2010_NBA_RGRC[1].pdf.

Lapchick, R., Kamke, C., and McMechan, D. (2009). The 2009 racial and gender report card: National Football League. Retrieved from http://www.tidesport.org/RGRC/2009/2009_NFL_RGRC.pdf.

Lassek, W.D., and Gaulin, S.J.C. (2009) Costs and benefits of fat-free muscle mass in men: Relationship to mating success, dietary requirements, and native immunity. *Evolution and Human Behavior* 30: 322–328.

Lehmann, L., and Feldman, M.W. (2008). War and the evolution of belligerence and bravery. *Proceedings of the Royal Society B* 275: 2877.

Lehrer, J. (2010, December 13). The truth wears off. *New Yorker.* Retrieved from http://www.newyorker.com/reporting/2010/12/13/101213fa_fact_lehrer?currentPage=all.

Lewontin, R.C. (1972). The apportionment of human diversity. *Evolutionary Biology* 6: 381–398.

———. (1983). Gene, organism and environment. In D.S. Bendall, ed., *Evolution from molecules to men,* pp. 273–285. Cambridge: Cambridge University Press.

Lindau, S.T., Schumm, L.P., Laumann, E.O., Levinson, W., O'Muircheartaigh, C.A., and Waite, L.J. (2007). A study of sexuality and health among older adults in the United States. *New England Journal of Medicine* 357(8): 762–774.

Long, J.C., Li, J., and Healy, M.E. (2009). Human DNA sequences: More variation and less race. *American Journal of Physical Anthropology* 139(1): 23–34.

Love and Rockets. (1987). "No new tale to tell." On *Earth, Sun, Moon* [CD]. London: Beggars Banquet Records.

Lovejoy, O. (2009). Reexamining human origins in light of *Ardipithecus ramidus. Science* 326: 108–115.

Maccoby, E.E., and Jacklin, C.N. (1974). *The psychology of sex differences.* Stanford, CA: Stanford University Press.

Madonna. (1989). "Express Yourself." On *Like a Prayer* [CD]. New York: Sire Records.

Madrigal, L., Blell, M., Ruiz, E., and Otárola-Durán, F. (2008). An evaluation of a genetic deterministic explanation for hypertension prevalence rate inequalities. In C. Panter-Brick and A. Fuentes, eds., *Health, Risk, and Adversity,* pp. 236–255. New York: Berghahn Press.

Malik, K. (2002). *Man, beast, and zombie: What science can and cannot tell us about human nature.* New Brunswick, NJ: Rutgers University Press.

Marks, J. (1995). *Human biodiversity: Genes, race, and history.* New York: Aldine de Gruyter.

———. (2000). Review of *Taboo* by Jon Entine. *Human Biology* 72: 1074–1078.

———. (2002). *What it means to be 98% chimpanzee.* Berkeley: University of California Press.

———. (2009). *Why I am not a scientist.* Berkeley: University of California Press.

Marshall, R.C. (2010). Introduction. In R.C. Marshall, ed., *Cooperation in social and economic life,* pp. vi–xxi. Lanham, MD: Altamira Press.

Mayr, E. (1963). *Animal speciation and evolution.* Cambridge, MA: Harvard University Press.

McKinley Jr., J.C. (2010, March 12). Texas conservatives win vote on textbook standards. *New York Times.* Retrieved from http://www.nytimes .com/2010/03/13/education/13texas.html.

Midgley, M. (2004). *The Myths We Live By,* 2nd ed. London: Routledge.

Miller, A., and Kanazawa, S. (2007, July 1). Ten politically incorrect truths about human nature. *Psychology Today* 40(4): 88–95.

Molnar, S. (2002). *Human variation: Races, types and ethnic groups.* Upper Saddle River, NJ: Prentice-Hall.

Montagu, A. (1942). *Man's most dangerous myth: The fallacy of race.* New York: Columbia University Press.

———. (1976). *The nature of human aggression.* Oxford: Oxford University Press.

Moses, Y. (2004). The continuing power of the concept of "race." *Anthropology and Education Quarterly* 35(1): 146–148.

Muller, M.N., and Wrangham, R.W. (2009). *Sexual coercion in primates and humans: An evolutionary perspective on male aggression against females.* Cambridge, MA: Harvard University Press.

Nanda, S. (2000). *Gender diversity: Cross-cultural variations.* Long Grove, IL: Waveland Press.

National Center for Health Statistics. (2010). *Health, United States, 2009: With special feature on medical technology.* Retrieved from http://www.cdc.gov/nchs/default.htm.

Nelson, R.J., and Chiavegatto, S. (2001). Molecular basis of aggression. *TRENDS in Neuroscience* 24(12): 713–719.

Nelson, R.J., and Trainor, B.C. (2007). Neural mechanisms of aggression. *Nature Reviews: Neuroscience* 8: 536–546.

Newman, T.K., Syagailo, Y.V., Barr, C.S., Wendland, J.R., Champoux, M., Grassle, M., Suomi, S.J., Higley, J.D., and Lesch, K. (2005). Monoamine oxidase A gene promoter variation and rearing experience influences aggressive behavior in rhesus monkeys. *Biological Psychiatry* 57: 167–172.

Nowack, M., and Highfield, R. (2011). *Super cooperators: Evolution, altruism and why we need each other to succeed.* New York: Free Press.

Nugent, H. (2007, October 17). Black people "less intelligent" scientist claims. *The Times.* Retrieved from http://www.timesonline.co.uk.

Odden, H.L. (2010). Interactions of temperament and culture: The organization of diversity in Samoan infancy. *Ethos* 37(2): 161–180.

Odling-Smee, F.J., Laland, K.N., and Feldman, M.W. (2003). *Niche construction: The neglected process of evolution.* Princeton, NJ: Princeton University Press.

Oka, R., and Fuentes, A. (2010). From reciprocity to trade: How cooperative infrastructures form the basis of human socioeconomic evolution. In R.C. Marshall, ed., *Cooperation in social and economic life,* pp. 3–28. Lanham, MD: Altamira Press.

Olmert, M. (2009). *Made for each other.* New York: Da Capo Press.

Ostrer, H. (2008). A genetic view of Jewish history. *Nature Genetics* 40(2): 127.

Ousley, S., Janytz, R., and Freid, D. (2009). Understanding race and human variation: Why forensic anthropologists are good at identifying race. *American Journal of Physical Anthropology* 139(1): 68–76.

Panter-Brick, C., and Fuentes, A. eds. (2008). *Health, risk, and adversity.* New York: Berghahn Press.

Parillo, V.N. (2009). *Strangers to these shores,* 9th ed. Boston: Allyn and Bacon.

Park, M.A. (2009). *Biological Anthropology,* 6th ed. New York: McGraw-Hill.

Pedersen, W.C., Miller, L.C., Putcha-Bhagavatula, A.D., and Yang, Y. (2002). Evolved sex differences in the number of partners desired? The long and the short of it. *Psychological Science* 13(2): 157–161.

Pinker, S. (2002). *The blank slate: The modern denial of human nature.* New York: Viking Press.

Pond, C.M. (1997). The biological origins of adipose tissue in humans. In M.E. Morbeck, A. Galloway, and l. Zihlman, eds., *The evolving female,* pp. 147–162. Princeton, NJ: Princeton University Press.

RACE. (1997). About the project: AAA's response to OMB directive 15. Retrieved from http://www.understandingrace.org/about/response.html.

———. (n.d.). Health connections: Sickle cell disease. Retrieved from http://www.understandingrace.org/humvar/sickle_01.html.

———. (n.d.). Sports quiz: White men can't jump and other beliefs about sports and race. Retrieved from http://www.understandingrace.org/lived/sports/index.html.

Race and Ethnicity. (n.d.). Retrieved from http://www.pollingreport.com/race.htm.

Reichard, U.H., and Boesch, C. eds. (2003). *Monogamy: Mating strategies and partnerships in birds, humans, and other mammals.* Cambridge: Cambridge University Press.

Relethford, J.H. (2002). Apportionment of global human genetic diversity based on craniometrics and skin color. *American Journal of Physical Anthropology* 118: 393–398.

Richards, R. (1987). *Darwin and the emergence of evolutionary theories of mind and behavior.* Chicago: University of Chicago Press.

Richerson, P.J., and Boyd, R. (2005). *Not by genes alone: How culture transformed human evolution.* Chicago: University of Chicago Press.

Rizolatti, G., and Sinigaglia, C. (2008). *Mirrors in the brain: How we share our actions and emotions.* Oxford: Oxford University Press.

Robinson, T.R. (2010). *Genetics for Dummies,* 2nd ed. Indianapolis, IN: Wiley Publishing.

Rotman, D.L. (2009). *Gendered lives: Historical archaeology of social relations in Deerfield, Massachusetts, ca. 1750–ca. 1904.* New York: Springer Academic Publishing.

Royal, D.C., Novembre, J., Fullerton, S.M., Goldstein, D.B., Long, J.C., Barnshad, M.J., and Clark, A.G. (2010). Inferring Genetic Ancestry: Opportunities, challenges, and implications. *American Journal of Human Genetics* 86: 661–673.

Ruff, C. (2002). Variation in human body size and shape. *Annual Reviews in Anthropology* 31: 211–232.

Rushton, J.P. (2000). *Race, evolution, behavior: A life history perspective,* 2nd ed. Port Huron, MI: Charles Darwin Research Institute Publications.

Ryan, C., and Jetha, C. (2010). *Sex at dawn: The prehistoric origins of modern sexuality.* New York: Harper Books.

Sanchez, R., Parkin, J.C., Chen, J.Y., and Gray, P.B. (2009). Oxytocin, vasopressin, and human social behavior. In P.T. Ellison and P.B. Gray, eds., *The endocrinology of social relationships,* pp. 319–339. Cambridge, MA: Harvard University Press.

Sapolsky, R. (1998). *Why zebras don't get ulcers.* New York: W.H. Freeman.

Schmitt, D.P. (2005). Sociosexuality from Argentina to Zimbabwe: A 48-nation study of sex, culture, and strategies of human mating. *Behavioral and Brain Sciences* 28: 247–311.

Shaver, P.R., and Mikulincer, M. (2011). Introduction. In P.R. Shaver and M. Mikulincer, eds., *Human aggression and violence: Causes, manifestations, and consequences,* pp. 123–142. Washington, DC: American Psychological Association.

Shepard, G.H., and Yu, D.W. (1998). Is beauty in the eye of the beholder? *Nature* 326: 321–322.

Shermer, M. (2007, January). Airborne baloney. *Scientific American* 296(1): 32.

Shweder, R.A. (2001). Rethinking the object of anthropology and ending up where Kroeber and Kluckhohn began. *American Anthropologist* 103(2): 437–440.

Siegel, A., and Victoroff, J. (2009). Understanding human aggression: New insights from neuroscience. *International Journal of Law and Psychiatry* 32: 209–215.

Smedley, A., and Smedley, B. (2005). Race as biology is fiction, racism as a social problem is real: Anthropological and historical perspectives on the social construction of race. *American Psychologist* 60(1): 16–26.

Smith, P. (2007, March 7). Ask the pilot. Retrieved from http://www.salon.com/technology/ask_the_pilot/2007/03/09/askthepilot224.

Smuts, B.B. (1987) Gender, aggression and influence. In Smuts, B.B., Cheney, D.L., Seyfarth, R.M., Wrangham, R.W., and Struhsaker, T.T., eds., *Primate Societies*, pp. 400–412. Chicago: University of Chicago Press.

Sprague, D. (1998). Age, dominance rank, natal status, and tenure among male macaques. *American Journal of Physical Anthropology* 105: 511–521.

Squire, S. (2008). *I don't: A contrarian history of marriage*. London: Bloomsbury Publishing.

Strier, K.B. (1994). Myth of the Typical Primate. *Yearbook of Physical Anthropology* 37: 233–71.

Stumpf, R.M. (2011). Chimpanzees and bonobos: Inter- and intraspecies diversity. In C. Campbell, A. Fuentes, K.C. MacKinnon, S. Bearder, and R. Stumpf, eds., *Primates in perspective*, 2nd ed., pp. 587–598. New York: Oxford University Press.

Summers, L. (2005, January 14). Remarks at the NBER conference. Retrieved from http://wiseli.engr.wisc.edu/archives/summers.php#conference-info.

Suomi, S. (2005). Genetic and environmental factors influencing the expression of impulsive aggression and serotinergic functioning in rhesus monkeys. In R. Tremblay, W.W. Hartup, and J. Archer, eds., *Developmental origins of aggression*, pp. 63–82. New York: Guilford.

Sussman, R.W. and Cloninger, C.R. (2012) *Cooperation and altruism*. New York: Springer.

Sussman, R.W., and Garber, P.A. (2011). Cooperation, collective action, and competition in primate social interactions. In C. Campbell, A. Fuentes, K.C. MacKinnon, S. Bearder, and R. Stumpf, eds., *Primates in perspective*, 2nd ed., pp. 587–598. New York: Oxford University Press.

Symons, D. (1979). *The evolution of human sexuality*. New York: Oxford University Press.

Tang-Martinez, Z. (2000). Paradigms and primates: Bateman's principle, passive females, and perspectives from other taxa. In S.C. Strum and L.M. Fedigan. eds., *Primate encounters: Models of science, gender, and society*, pp. 261–274. Chicago: University of Chicago Press.

Tattersall, I. (1997). *The fossil trail: How we know what we think we know about human evolution.* New York: Oxford University Press.

Templeton, A. (1999). Human races: A genetic and evolutionary perspective. *American Anthropologist* 100: 632–650.

They Might Be Giants. (1990). "Your Racist Friend." On *Flood* [CD]. New York: Elektra/Asylum Records.

Trivers, R.L. (1972). Parental investment and sexual selection. In B. Campbell, ed., *Sexual selection and the descent of man 1871–1971*, pp. 136–179. Chicago: Aldine.

Tylor, E.B. (1871). *Primitive culture.* London: John Murray.

UN General Assembly. (1948). The universal declaration of human rights. Retrieved from http://www.un.org/en/documents/udhr/index.shtml.

US Bureau of Labor Statistics. (2009). Labor force characteristics by race and ethnicity, 2008 (Report 1020). Retrieved from http://www.bls.gov/cps/cpsrace2008.pdf.

US Census Bureau, (2010). 2010 census questionnaire reference book. Retrieved from http://2010.census.gov/partners/pdf/langfiles/qrb_English.pdf.

Waddington, C.H. (1959). Evolutionary adaptation. In S Jax, ed., *Evolution after Darwin*, pp. 381–402. Chicago: University of Chicago Press.

Wade, N. (2007, April 10). Pas de deux of sexuality is written in the genes. *New York Times.* Retrieved from http://www.nytimes.com/2007/04/10/health/10gene.html.

———. (2010, July 19). Adventures in very recent evolution. *New York Times.* Retrieved from http://www.nytimes.com/2010/07/20/science/20adapt.html.

Walker, P.L. (2001). A bioarcheological perspective on the history of violence. *Annual Reviews in Anthropology* 30: 573–596.

Watts, D.P., Muller, M., Amsler, S.J., Mbabazi, G., and Mitani, J.C. (2006). Lethal intergroup aggression by chimpanzees in Kibale National Park, Uganda. *American Journal of Primatology* 68: 161–180.

Wedekind, C., and Penn, D. (2000). MHC genes, body odours and odour preferences. *Nephrology Dialysis Transplantation* 15: 1269–1271.

Weiss, K.M. (1998). Coming to terms with human variation. *Annual Reviews in Anthropology* 27: 273–300.

———. (2008). The good, the bad, and the ugli: Can seemingly incompatible worldviews be hybridized? *Evolutionary Anthropology* 17: 129–134.

Weiss, K.M, and Buchanan, A.V. (2009). *The Mermaid's tail: Four million years of cooperation in the making of living things.* Cambridge, MA: Harvard University Press.

Wheatley, T., Milleville, S.C., and Martin, A. (2007). Understanding animate agents: Distinct roles for the social network and mirror system. *Psychological Science* 18: 469.

White, E.J. (2011, January 4). The n-word belongs in "Huckleberry Finn." Retrieved from http://www.salon.com/life/this_week_in_blackness/2011/01/04/huck_finn_n_word.

Williams, M.E. (2010, August 13). Dr. Laura's n-bomb meltdown. Retrieved from http://www.salon.com/life/feature/2010/08/13/dr_laura_the_n_word.

Wilson, M., and Daly, R. (2009). Coercive violence by human males against their female partners. In M.N. Muller and R.W. Wrangham, eds., *Sexual coercion in primates and humans: An evolutionary perspective on male aggression against females*, pp. 319–339. Cambridge, MA: Harvard University Press.

Wood, J.L., Heitmiller, D., Andreasen, N.C., and Nopoulos, P. (2008). Morphology of the ventral frontal cortex: Relationship to femininity and social cognition. *Cerebral Cortex* 18: 534–540.

Wood, J.L., Murko, V., and Nopoulos, P. (2008). Ventral front cortex in children: Morphology, social cognition and femininity/masculinity. *SCAN* 3: 168–176.

Wood, W., and Eagly, A.H. (2002). A cross-cultural analysis of the behavior of women and men: Implications for the origins of sex differences. *Psychological Bulletin* 128(5): 699–727.

Wrangham, R. (2010, April 2). Why we kill. Retrieved from http://bigthink.com/ideas/19361.

Wrangham, R., and Peterson, D. (1996). *Demonic males: Apes and the origin of human violence*. New York: Mariner Books.

Wulf, S. (1987, February 9). Standing tall at short. *Sports Illustrated* 66(6): 132–148.

Index

TEXT
10/13 Sabon

DISPLAY
Din

COMPOSITOR
Toppan Best-set Premedia Limited

PRINTER AND BINDER
Maple-Vail Book Manufacturing Group